Proceedings in Engineering Mechanics

Research, Technology and Education

Series Editors

Lucas F. M. da Silva, Faculty of Engineering, University of Porto, Porto, Portugal

António J. M. Ferreira ⓘ, Faculty of Engineering, University of Porto, Porto, Portugal

This book series publishes the results of meetings dealing with material properties in engineering and science. It covers a wide range of topics, from the fundamentals of materials mechanics and applications for various industries to aspects of scientific training and career development. The volumes in the series are based typically on primary research materials presented at conferences, workshops, and similar scientific meetings, and represent comprehensive scientific and technical studies.

Lucas F. M. da Silva · Digavalli Ravi Kumar ·
Maria de Fátima Reis Vaz · Ricardo J. C. Carbas
Editors

1st International Conference on Engineering Manufacture 2022

Selected Contributions of EM 2022

 Springer

Editors
Lucas F. M. da Silva
Faculty of Engineering
University of Porto
Porto, Portugal

Maria de Fátima Reis Vaz
Instituto Superior Técnico
University of Lisbon
Lisbon, Portugal

Digavalli Ravi Kumar
Department of Mechanical Engineering
Indian Institute of Technology Delhi
New Delhi, India

Ricardo J. C. Carbas
Institute of Science and Innovation
in Mechanical and Industrial Engineering
(INEGI)
Porto, Portugal

ISSN 2731-0221   ISSN 2731-023X   (electronic)
Proceedings in Engineering Mechanics
ISBN 978-3-031-13233-9   ISBN 978-3-031-13234-6   (eBook)
https://doi.org/10.1007/978-3-031-13234-6

Preface

This volume of *Proceedings in Engineering Mechanics—Research, Technology and Education* contains selected papers presented at the 1st International Conference on Engineering Manufacture 2022 (EM 2022), held in Porto (Portugal) during 5–6 May 2022.

This conference is held every 2 years. The conference is chaired by Lucas F. M. da Silva (University of Porto, Portugal) and co-chaired by Murat Tiryakioğlu (Jacksonville University, Florida, USA), Digavalli Ravi Kumar (Indian Institute of Technology Delhi, India), Maria de Fátima Reis Vaz (University of Lisbon, Portugal), Ricardo J. C. Carbas (INEGI, Portugal) and Eduardo Marques (INEGI, Portugal). The focus is on engineering manufacture and includes works on additive manufacturing, precision machining, vacuum-assisted high-pressure die casting, semi-solid metal casting, and compressive and tensile forming processes, among many others. Special topics of interest are

- modeling and optimization of manufacture processes
- design for manufacturing strategies
- development of new manufacturing technologies
- design of novel manufacturing equipment
- comparative case studies
- cost and quality analysis
- ecological aspects
- destructive and non-destructive testing of manufactured components.

69 abstracts were presented, representing 12 countries. Portugal and Germany are the most represented countries. The main themes treated are Additive manufacturing, Forming, Joining, Machining, Molding and Casting and Manufacturing equipment and maintenance.

In order to disseminate the work presented at EM 2022, selected papers were prepared which resulted in the present volume. A wide range of topics are covered resulting in nine chapters dealing with additive manufacturing (first three chapters), machining (1 chapter), injection process (1 chapter), joining (2 chapters) and applications (2 chapters). The book provides a state of the art of engineering manufacture and also serves as a reference volume for researchers and graduate students using technological processes.

The organizer and editor wish to thank all the authors for their participation and cooperation, which made this volume possible. Finally, I would like to thank the team of Springer-Verlag, especially Dr. Christoph Baumann and Ute Heuser, for the excellent cooperation during the preparation of this volume.

Porto, Portugal   Lucas F. M. da Silva
                                                                                         lucas@fe.up.pt

New Delhi, India   Digavalli Ravi Kumar
Lisbon, Portugal   Maria de Fátima Reis Vaz
Porto, Portugal   Ricardo J. C. Carbas
June 2022

Contents

Applications

Additive Manufacturing

Mechanical Assessment of PBF-EB Manufactured IN718 Lattice Structures

Daniel Kotzem⑩ **and Frank Walther**⑩

Abstract Based on the layer-by-layer manufacturing, additive manufacturing techniques offer numerous possibilities for the design of new lightweight constructions. On the way to a reliable application within safety-relevant components, the deformation and damage behavior, especially under cyclic loading, have to be understood in detail. Within this investigation, the mechanical behavior of electron beam powder bed fusion (PBF-EB) Inconel 718 (IN718) F_2CC_Z lattice structures is characterized. Computer tomographic scans are conducted in order to quantify the present defect density as well as the nominal cross section. Overall porosity was found to be low, however, as-built structure shows significant deviations in terms of geometrical accuracy compared to the CAD model. The fatigue behavior is investigated by means of coupled multiple (MAT) and constant amplitude tests (CAT) at fully-reversed loading using various measuring techniques such as digital image correlation (DIC) and direct current potential drop system (DCPD). Results from MAT are in good agreement with CAT since load levels for 10^4 and $2 \cdot 10^6$ cycles (run out) could be detected based on the material responses. A run out could be realized at $\sigma_a = 35$ MPa, however, process-induced surface roughness is the main reason for lower fatigue performance due to local formation of stress concentrations and multiple crack initiation sites. During cyclic loading, particular material reactions were identified, consisting of partial failure of single struts and, as a consequence, of several linked unit cells which could be correlated with the measurement quantities recorded.

Keywords Additive manufacturing · Electron beam powder bed fusion
(PBF-EB) · Ni-based alloys · Lattice structures · Damage tolerance

D. Kotzem (✉) · F. Walther
Chair of Materials Test Engineering, TU Dortmund University, Baroper Str. 303, 44227 Dortmund, Germany
e-mail: daniel.kotzem@tu-dortmund.de

F. Walther
e-mail: frank.walther@tu-dortmund.de

© The Author(s), under exclusive license to Springer Nature Switzerland AG 2023
L. F. M. da Silva et al. (eds.), *1st International Conference on Engineering Manufacture 2022*, Proceedings in Engineering Mechanics,
https://doi.org/10.1007/978-3-031-13234-6_1

1 Introduction

Additive manufacturing (AM) offers the possibility to design new lightweight components especially in the aerospace industry to fulfill upcoming requirements aiming on increasing efficiency and sustainability. In a recent case study, SLM Solutions and Cellcore have developed a rocket nozzle with integrated cooling channels based on lattice structures [1]. Base material for the rocket nozzle is the Ni-based alloy Inconel®718 (IN718). IN718 is known as a precipitation hardenable alloy with superior mechanical properties even at elevated temperatures up to 650 °C and good chemical resistance [2]. However, due to excellent mechanical, physical and thermal properties machining of IN718 is challenging and cost-intensive [3]. Various researchers [4, 5] have already processed IN718 with powder bed fusion processes (PBF) such as laser (PBF-LB/M) and electron beam powder bed fusion (PBF-EB). Within the PBF process, a thin powder layer is spread over the build platform. A high energy source (laser or electron beam) is used to locally melt the powder particles according to the computer-aided designed (CAD) model. Subsequently, the build platform is lowered and a new powder layer is applied. The aforementioned steps are repeated until the part is completed [6]. It could be demonstrated that PBF manufactured IN718 can have similar or even enhanced mechanical properties compared to wrought material [4]. Overall aim might be combining the near-net shape fabrication with the outstanding design flexibility of the PBF manufacturing techniques for future applications which can lead to significant time and cost savings [7].

Considering complex structures, especially periodic lattice structures are currently focused since their mechanical properties can easily be tailored by varying the lattice type and corresponding dimensions (e.g. strut diameter and unit cell size) compared to foams and honeycombs [8–10]. Generally, the deformation behavior of lattice structures can be either stretch- or bending-dominated and can be quantified by Maxwell's stability criterion [11]. Based on the deformation behavior, the lattice structures exhibit different mechanical properties. Typically, stretch-dominated lattice structures are preferred for lightweight applications due to higher specific strength and stiffness, however, bending-dominated lattice structures are more compliant and are used in applications where good absorbing and damping properties are needed [12]. The PBF-EB process is predestined to manufacture lattice structures due to the sintering effect of the powder bed which make support structures nearly unneeded [13]. In literature, several studies were conducted to investigate the influence of different process parameters [14], cell morphologies [15], relative densities [16], heat treatments [17] and build orientations [14] on the mechanical behavior of PBF-EB manufactured lattice structures. Thereby, quasi-static compression and tensile tests are the preferred choice to characterize the mechanical behavior. Only few studies focused on highlighting the fatigue behavior including the detection of deformation and damage mechanisms. Zhao et al. [18] investigated the influence of different cell types on the compression fatigue behavior. More recently, fatigue tests for PBF-EB manufactured Ti6Al4V lattice structures were conducted by

Lietaert et al. [19] for different fatigue regimes such as tension-tension, compression-compression and tension–compression. Due to the influence of local mean stress, compression-compression and tension-tension fatigue loading leads to early fatigue failure compared to fully-reversed loading. Regarding the fatigue behavior of PBF manufactured IN718 lattice structures, very few studies can be found. Solely, Huynh et al. [20] investigated the fatigue behavior of such lattices. It was shown that failure of the structures was located in the intersection between nodes and struts which was in line with the corresponding finite element analysis. Further on, the damage progress was highlighted consisting of failure of a first strut and subsequent formation of new stress concentrations at surrounding struts. Thereby, struts are loaded under compression, tension and shear [21]. As a result, subsequent failure takes place and the previous steps are repeated until catastrophic failure occurs. The high surface-to-volume ratio, in combination with the as-built surface roughness, is expected to act as a catalyst for early crack initiation and propagation due to the presence of many notch-like defects at the surface ending up in multiple crack initiation sites.

To sum up, many factors such as lattice type and dimensions, process parameters, surface quality and defect state can influence the mechanical properties of PBF-EB manufactured lattice structures and the deformation and damage behavior are yet not fully analyzed. To address this, additional measurement techniques were adapted and implemented by the author's in prior investigations [22–24] to correlate the measurement quantities with particular material reactions. The region of interest was gradually extended beginning with monitoring the deformation and damage behavior of single struts, single unit cell planes and complete lattice structures consisting of several linked unit cells. Within this work, the mechanical behavior of PBF-EB manufactured IN718 lattice structures is characterized. For the lattice structure, the stretch-dominated F_2CC_Z lattice type was selected. Primary investigations include microstructural analysis and computer tomographic scans to quantify the present defect density and the nominal cross section. The fatigue behavior is investigated by means of coupled multiple and constant amplitude tests at fully-reversed loading using digital image correlation (DIC) and direct current potential drop system (DCPD) to capture the deformation and damage progress.

2 Experimental Procedure

The Inconel®718 alloy (IN718) was processed by means of PBF-EB using an A2X system (Arcam AB, Moelndal, Sweden). The manufacturing process consists of the typical steps: pre-heating, contour and hatch melting, post-heating, lowering build platform and recoating with a new powder layer. The process parameters used are listed in Table 1.

The specimens were manufactured upright, i.e. parallel to building direction. The F_2CC_Z lattice type was used for the lattice structures and the initial strut diameter was 1.2 mm. The unit cell size was set to $7 \times 7 \times 7$ mm^3 leading to a relative density of 0.1263. In total, $3 \times 3 \times 3$-unit cells form the lattice structure inside

Table 1 Process parameters used to manufacture the IN718 specimens by means of electron beam powder bed fusion (PBF-EB)

Volume energy	Beam speed	Beam current	Hatch distance	Focus offset	Layer thickness
25 J/mm^2	2,500 mm/s	12.5 mA	160 μm	−3 mA	75 μm

the specimen. After manufacturing, specimens were machined to create the threads which were used to mount the specimen for later mechanical testing. No additional heat treatments were conducted and the specimen were tested in as-built state. The final specimen geometry is shown in Fig. 2c.

Subsequently, specimens were cold embedded, ground with SiC and polished with diamond suspension. To highlight the present microstructure, the polished surface was etched for 30 s at 60 °C using the "Beraha II" etching solution. For analyzing, the light microscope Axio Imager M1 (Carl Zeiss, Goettingen, Germany) was selected. In a next step, computer tomographic scans were conducted using the XT-H 160 system (Nikon Metrology, Tokyo, Japan), whereby a minimal resolution of 31 μm could be achieved. The detailed scanning parameters are listed in Table 2.

Based on the reconstructed volumes, the defect state was quantitatively analyzed by the software VGStudio Max 2.2 (Volume Graphics, Heidelberg, Germany) using the algorithm "VGDefX (2.2)". Additionally, the following steps were conducted for every specimen to determine the nominal cross section: (a) Slices at positions with lowest cross section were determined (cf. Fig. 1a); (b) every slice was imported and scaled using the software ImageJ (cf. Fig. 1b); (c) the image was converted into a black/white image (binarization) based on a defined threshold value (cf. Fig. 1c); (d) the nominal cross section could be calculated for every specimen (cf. Fig. 1d). For each build job, at least three specimens were penetrated by X-ray, respectively.

Table 2 Scanning parameters for computed tomography scans (μCT)

Material	Beam energy	Beam current	Power	Effective pixel size	Exposure rate	
IN718	140 kV	66 μA	9.2 W	31 μm	354 ms	2.82 fps

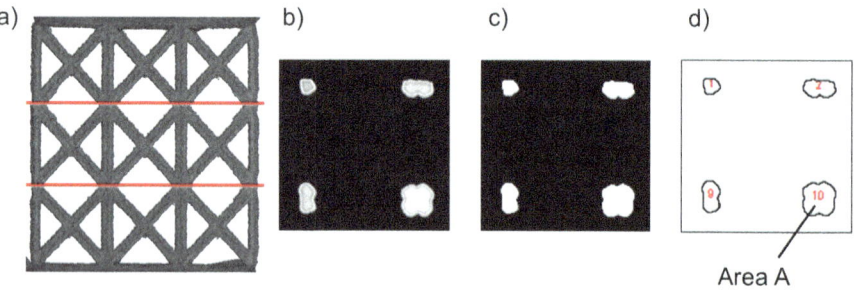

Fig. 1 Processing steps to calculate the nominal cross section; **a** schematic representation of the cutting planes, **b** extracted image, **c** binarization and **d** analyzed data

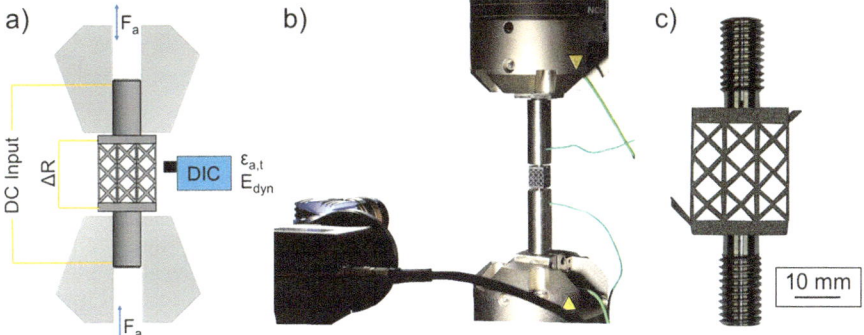

Fig. 2 a Schematic illustration and **b** experimental setup used for the cyclic testing; **c** specimen geometry

In respect of the mechanical characterization, fatigue tests were conducted using a combination of multiple (MAT) and constant amplitude tests (CAT). Therefore, the servohydraulic system Schenck PC63M with Instron 8800 controller equipped with a 45 kN load cell was selected. All tests were carried out at a load ratio of $R = -1$ (fully-reversed loading). The test frequency was $f = 5$ Hz and the initial load for the MAT was set to $\sigma_a = 5$ MPa and was increased stepwise by 5 MPa after every 10^4 cycles. Material reactions were captured using an application-specific instrumentation which was previously introduced in [24]. It is composed of a direct current potential drop (DCPD) measurement system (Ametek, Berwyn, PA, USA) and contactless digital image correlation (DIC) using the Imager M-Lite camera (LaVision, Goettingen, Germany). To detect local and integral strain distributions, specimens were grounded black and a white speckle-pattern was applied. The schematic illustration as well as the experimental setup can be seen in Fig. 2a, b. Due to relatively long test durations, a triggered image acquisition was chosen [23]. The spacing between two trigger points was adjusted as a function of the assumed fatigue life. Based on the data, the total strain amplitude $\varepsilon_{a,t}$ was calculated. To detect stiffness changes during the tests, dynamic Young's modulus E_{dyn} was even calculated whereby E_{dyn} was approximated by the gradient of a linear slope between the two reversal points of the hysteresis loop.

3   Results and Discussion

3.1   Microstructure and μCT Analysis

The as-built microstructure parallel to building direction for the F_2CC_Z lattice structure is shown in Fig. 3, whereby the intersection from bulk material to small-scale strut as well as a central nodal point are highlighted. As can be seen, predominant columnar oriented grains with strong texture along building direction (BD) are visible which is in line with known literature [25, 26]. Further on, grains grow over several material layers implicating epitaxial growth [27]. Smaller equiaxed grains can be found in the contour region which result from aggregation of unmelted or partially melted powder particles [28]. Considering the microstructure inside the nodal point in the center of the lattice structure, it can be highlighted that columnar oriented grains even show a strong texture along BD although struts were built in 45° orientation which was earlier proposed by Sun et al. [28]. Smaller gas pores can be detected within the light microscope images, however, μCT analysis revealed a relatively low porosity, whereby a total volume of $1.23 \cdot 10^{12}$ μm^3 was analyzed. It has to be mentioned that smallest resolution which could be realized by μCT was 31 μm due to the size of the lattice specimen.

To further analyze the pore distribution, pore density and sphericity, which is a measure of how spherical an object or, in this particular case, a defect is [29], are calculated and results are plotted in Fig. 4a, b. Additionally, the corresponding reconstructed volume is shown in front and top view (cf. Fig. 4c). In general, porosity can be found almost exclusively in vertical struts which are located in the center of the specimen. Most defects have an equivalent pore diameter ranging between 200 and 300 μm. Only two pores exhibited an equivalent pore diameter > 800 μm. The shape of the pores tends to be more elongated than round since average sphericity was found to be 0.53.

Generally, internal defects might be either gas pores, lack of fusion (LoF) or shrinkage porosity and mainly result from non-optimal process parameters [30]. However, gas pores were found to result from gas entrapments inside the powder particles and are randomly distributed within the volume [31]. More detrimental for

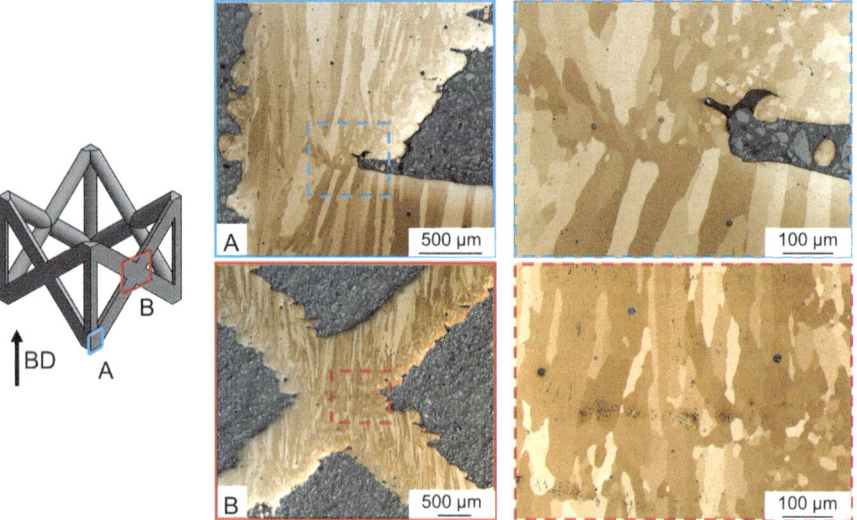

Fig. 3 As-built microstructure of PBF-EB manufactured IN718 lattice structure

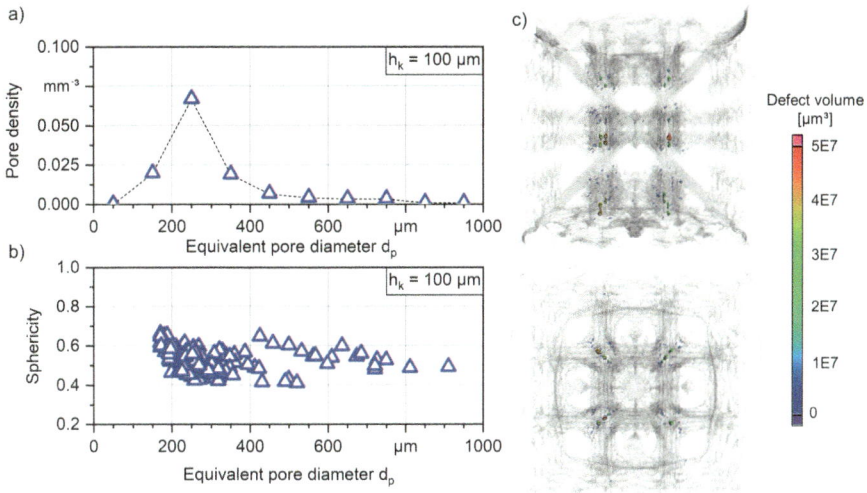

Fig. 4 μCT analysis of an exemplary PBF-EB manufactured IN718 F_2CC_Z lattice structure: **a** pore density; **b** sphericity; **c** reconstructed volume in front and top view

the cyclic properties are LoF defects having a more elongated shape which leads to the formation of local stress concentrations. They are preferably located between two material layers and are caused due to insufficient melting [32]. Especially the occurrence of LoF defects at the near-surface and the process-induced surface roughness can decrease the fatigue performance significantly [5].

Taking into account the shape of as-built PBF structures, it was already shown in [33, 34] that these structures can exhibit alternating geometries compared to the CAD model. To enable a reliable stress calculation for later mechanical testing, μCT scans were conducted to determine the nominal cross section of the specimens. For every build job, at least three specimens are penetrated by X-ray, respectively. The detailed approach is summed up in Sect. 2. The value for the nominal cross section which was calculated based on the μCT scans as well as the CAD cross section are listed in Table 3.

Comparing the cross sections, significant deviations can be noticed. In total, an average cross section of 24.45 ± 2.97 mm² could be determined for the F_2CC_Z specimens leading to a positive deviation of around 39% compared to the CAD model. In literature both negative and positive deviations were reported for PBF manufactured structures [33, 35]. The angle of an individual strut towards the building plane was found to be one main factor for geometrical deviations [36]. Especially

Table 3 Calculated nominal cross section for fatigue specimens based on CAD and μCT data	Lattice type	F_2CC_Z
	CAD cross section	17.50 mm²
	μCT cross section	24.45 ± 2.97 mm²

horizontal or oblique oriented struts show a more elliptical shape due to powder adhesions on the down-facing side which can increase the surface roughness as well [37]. Van Bael et al. [38] proposed that increased strut thicknesses as well as surface irregularities can even result from the staircase effect which is also present in AM parts. Within this investigations, nominal cross section was calculated at locations with presumed smallest cross section. However, these locations are always in the nodal point where oblique and vertical oriented struts join together. As shown in [16, 36] powder particles can agglomerate within pores or corners in a lattice structure which might explain the positive deviations compared to the ideal CAD cross section. As a consequence, different approaches are currently focused to consider geometrical mismatches between CAD and as-built components in order to facilitate the design and construction process [39].

3.2   Mechanical Properties

Fatigue tests were conducted to characterize the mechanical behavior of PBF-EB manufactured IN718 F_2CC_Z lattice structures. To estimate the fatigue life in a time- and cost-efficient manner, a combination of multiple (MAT) and constant amplitude tests (CAT) was used [40]. The results of the MAT are shown in Fig. 5, whereby stress amplitudes (σ_a) are colored in black, changes in electrical resistance (ΔR_{DC}) in orange and the total strain amplitude ($\varepsilon_{a,t}$) in blue. A maximum stress amplitude of $\sigma_a = 100$ MPa with a corresponding number of cycles to failure $N_f = 165,117$ cycles was achieved by the F_2CC_Z lattice structure. At an early stage of the test, a slight linear increase can be detected for $\varepsilon_{a,t}$ at the load step from 35 to 40 MPa. With increasing number of cycles, the linear course changes to an exponential growth and final failure occurred shortly after. For ΔR_{DC}, a first reaction can be detected at the load level of $\sigma_a = 75$ MPa. The subsequent course is comparable to the results of $\varepsilon_{a,t}$, however, sudden increases in ΔR_{DC} can be detected which can be linked to local failure of single struts [22]. The range of first material response and final failure defines the stress levels for the subsequent CAT which were selected between 25 and 105 MPa.

In contrast to the damage progress of bulk material, lattice structures show a more complex deformation and damage behavior under quasi-static and cyclic loading. To fully analyze the damage progress, an exemplary CAT with corresponding material responses for a specimen which was tested at $\sigma_a = 95$ MPa ($N_f = 60,285$ cycles) is shown in Fig. 6. Coincident to the MAT, ΔR_{DC} is colored in orange and $\varepsilon_{a,t}$ in blue. Based on the collected data, dynamic Young's modulus E_{dyn} as well as the ratio between dynamic Young's modulus under compression ($E_{dyn, comp.}$) and tension ($E_{dyn, ten.}$) is plotted versus the number of cycles N. Specific material reactions could be determined within the CAT and corresponding DIC images are named as A–D (cf. Figure 6). At the beginning of the test (stage A) the specimen is shown in damage-free condition, however, from the beginning a linear increase can be detected for $\varepsilon_{a,t}$ and ΔR_{DC} which can be attributed to early crack initiation and propagation due

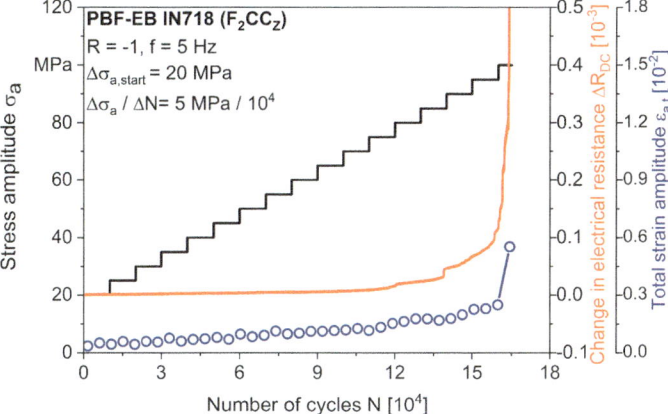

Fig. 5 Material responses recorded to be used for the analysis of deformation and damage progress of F_2CC_Z lattice structures in the multiple amplitude test

to the increased load level [41]. At stage B, first partial failure within the lattice structure occurred and is located at an outer strut on the top right side. Comparing the material responses, a sudden increase in ΔR_{DC} as well as a stiffness degradation can be detected which is more pronounced with increasing number of cycles until catastrophic failure. Between stage B and C, another sudden increase in ΔR_{DC} can be detected which implicates further partial failure within the lattice, however, this could not be captured by the DIC. Before reaching stage C, the course of all measurement responses changes from linear to exponential growth. At stage C and D, further local failure within the structure occurred in a plane perpendicular to load direction. As can be seen, partial failure at stage D takes place in direct surrounding to previous failure (cf. stage C). Considering the ratio between $E_{dyn, comp.}$ and $E_{dyn, ten.}$ for the entire test, a constant course is visible leading to the assumption that no buckling of the hysteresis loop takes place [23]. The results for all CAT are shown in Fig. 7 in a S–N curve. As can be seen, approx. 10^4 cycles could be reached by specimens which were tested at $\sigma_a = 105$ MPa and a run-out ($2 \cdot 10^6$ cycles) could be realized at $\sigma_a = 35$ MPa. The results are in good agreement with the MAT which might enable the reduction of fatigue tests for lattice structures in future investigations. To further describe the fatigue life, Basquin equation was used [42]. The corresponding equation is highlighted in Fig. 7 and gives a good approximation with a coefficient of determination $R^2 = 0.79$.

Material reactions from bulk material are typically captured by means of total strain amplitude and dynamic Young's modulus. Supplementary, Piotrowski et al. [41] proposed that resistance measurement techniques can be used as well and can be linked to the actual defect state or rather the remaining cross section of the component [43]. However, lattice structures show a more complex deformation and damage behavior which is comparable to metallic foams [44]. Authors have previously shown in [22, 23] that the damage progress can be described by in-situ

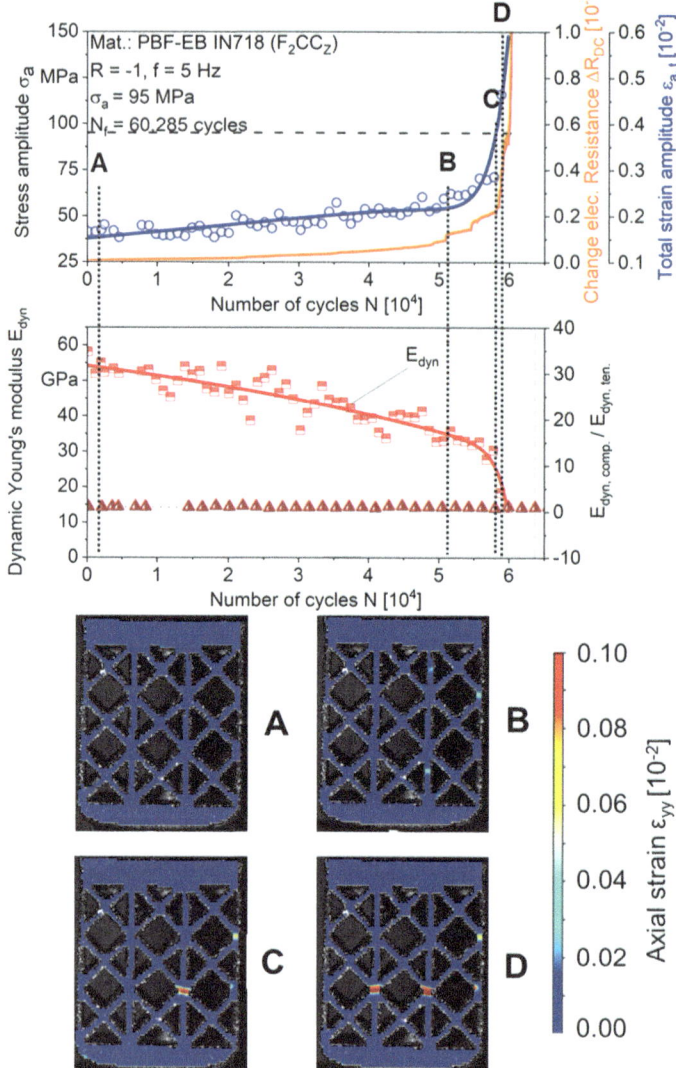

Fig. 6 Material responses recorded for a F_2CC_Z lattice structure ($\sigma_a = 95$ MPa, $N_f = 60{,}285$ cycles) to evaluate the damage progress in constant amplitude tests with corresponding digital image correlation images at specific stages (A-D)

measurement techniques such as DIC, thermography and DCPD. In a recent study, Radlof et al. [45] even used supplementary measurement techniques to characterize the fatigue behavior under cyclic bending and torsion. As mentioned in [46], lattice structures undergo three stages during unidirectional fatigue loading. Within stage I,

Fig. 7 S–N curve for the
PBF-EB IN718 F_2CC_Z
lattice structures

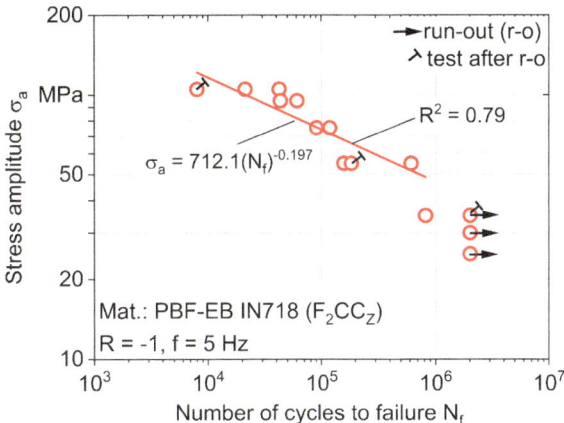

the structures are dominated by cyclic creep which leads to a progressive accumulation of plastic strain. First crack initiation occurs in stage II [47]. With increasing number of cycles crack propagation leads to first partial failure within the lattice structure which defines the transition to stage III. Sudden strain increases were reported by Sugimura et al. [48] within stage III and were attributed to local failure. As a consequence, multiple failure within several linked unit cells is present ending in final failure [49]. Based on the results, the use of DIC and DCPD can help to identify stage II and III, respectively. In detail, a linear increase in $\varepsilon_{a,t}$ and ΔR_{DC} as well as a decrease in E_{dyn} were noticed from the beginning of the test implicating that stage I was skipped due to early crack initiation. Based on literature, predominant location for crack initiation was found at the specimens' surface due to high surface roughness and corresponding notch-like defects [50]. Later on, linear course of all measurement techniques changes to exponential growth clearly highlighting the transition to stage III. Especially in the transition from stage II to III, single strut or rather failure in nodal points could reliable be detected based on sudden increases in ΔR_{DC} (cf. Fig. 6). Especially imaging techniques are the preferred choice to detect single strut failure, however, increasing complexity of the component makes it challenging to detect all material reactions [46].

Considering the literature, it is quite complicated to compare the mechanical properties of PBF manufactured lattice structures since influencing factors such as process parameters, lattice type, relative density and defect state are present. Further on, missing standards for a specimen design and the mechanical characterization are additional obstacles. Only for quasi-static compression testing, ISO 13314 [51] delivers a standard for a potential specimen design, however, they are not suitable for fully-reversed loading (R $= -1$), thus, new standards have to be introduced. As proposed by Benedetti et al. [49], AM lattice structures tend to early failure during fatigue loading due to several reasons. A major aspect is the cell morphology since bulk material is substituted by a unit cell consisting of struts which are connected at nodal points. Due to the individual connectivity, a reduced load bearing area is present and local stress concentrations can arise. Another concern might be geometrical inaccuracies such as size deviations and as-built surface roughness compared to the CAD model [35, 49]. Especially the angle between single struts relative to the building direction is critical, contributing to a high surface roughness. Further on, Razavi et al. [52] have stated that deep micro-notches, which typically result from the as-built roughness, reduce more the effective cross section of small-scale structures than in bulk material clearly addressing post-processing techniques which are needed to enhance the fatigue performance of AM lattice structures. Future investigations will address this issue by qualifying potential post-processing methods such as electrochemical polishing to improve the surface roughness and, thus, the fatigue performance. On the other hand, intermittent fatigue tests in combination with in-situ µCT loading stage are intended to fully understand and describe the three-dimensional damage progress of AM lattice structures which might provide the fundament for fatigue life prediction.

4   Conclusions and Outlook

Additive manufacturing processes such as laser (PBF-LB/M) and electron beam powder bed fusion (PBF-EB) are more and more frequently used in industrial applications due to advanced lightweight design with nearly no geometrical limitations. To enable an application, especially in safety-relevant components, the underlying deformation and damage mechanisms of complex structures have to be known. Within this work, the mechanical behavior of PBF-EB manufactured F_2CC_Z lattice structures made from Inconel 718 was investigated. Based on computer tomographic scans, a low porosity was detected, however, geometrical deviations compared to the CAD model are present. In particular, positive deviations in the range of 40% could be determined at nodal points, most probably resulting from powder agglomerations. Within cyclic tests under fully-reversed loading, the fatigue behavior was characterized using digital image correlation (DIC) and direct current potential drop system (DCPD). In detail, results from multiple amplitude test are in good agreement with constant amplitude tests in the range between 10^4 and $2 \cdot 10^6$ cycles. A run out could be realized at $\sigma_a = 35$ MPa. Fatigue life could be estimated based on Basquin equation with a coefficient of determination $R^2 = 0.79$. Additionally, particular material reactions such as single strut failure could reliably correlated with the measurement quantities captured.

Future investigations will focus on improving the as-built surface roughness since notch-like defects at the surface were found to be preferred crack initiation sites. Further on, intermittent fatigue tests are conducted whereby the three-dimensional damage progress will be determined by means of in-situ µCT stage to enable a later fatigue life prediction.

Acknowledgements   The authors thank the German Research Foundation (Deutsche Forschungsgemeinschaft, DFG) for its financial support within the research projects No. 379213719 "Damage tolerance evaluation of electron beam melted cellular structures by advanced characterization techniques" (NI 1327/13-1, WA 1672/32-1) and "Microstructure- and defect-controlled damage tolerance evaluation of lattice structures at room temperature and 650 °C based on the E-PBF processed Ni-based alloy Inconel 718 (NI 1327/13-2, WA 1672/32-2). Furthermore, authors would like to thank Thomas Niendorf and Tizian Arold (University of Kassel) for providing the investigated material in the framework of an excellent scientific collaboration.

References

1. Donath, S.: 3D printing a rocket engine. https://www.etmm-online.com/3d-printing-a-rocket-engine-a-886960/. Accessed 23 May 2022
2. Dehmas, M., Lacaze, J., Niang, A., Viguier, B.: TEM study of high-temperature precipitation of delta phase in Inconel 718 alloy. Adv. Mater. Sci. Eng. 1–9 (2011). https://doi.org/10.1155/2011/940634
3. Uhlmann, E., Wiemann, E., Zettier, R.: Untersuchung des Zerspanverhaltens von Inconel 718. wt Werkstattstechnik online **95**(1/2), 62–67 (2005)
4. Sanchez, S., Smith, P., Xu, Z., Gaspard, G., Hyde, C.J., Wits, W.W., Ashcroft, I.A., Chen, H., Clare, A.T.: Powder bed fusion of nickel-based superalloys: a review. Int. J. Mach. Tools Manuf. **165**, 103729 (2021). https://doi.org/10.1016/j.ijmachtools.2021.103729
5. Balachandramurthi, A. R., Moverare, J., Mahade, S., Pederson, R.: Additive manufacturing of alloy 718 via electron beam melting: effect of post-treatment on the microstructure and the mechanical properties. Materials (Basel, Switzerland) **12**(1) (2018). https://doi.org/10.3390/ma12010068
6. Singh, R., Gupta, A., Tripathi, O., Srivastava, S., Singh, B., Awasthi, A., Rajput, S.K., Sonia, P., Singhal, P., Saxena, K.K.: Powder bed fusion process in additive manufacturing: an overview. Mater. Today: Proc. **26**, 3058–3070 (2020). https://doi.org/10.1016/j.matpr.2020.02.635
7. Blakey-Milner, B., Gradl, P., Snedden, G., Brooks, M., Pitot, J., Lopez, E., Leary, M., Berto, F., Du Plessis, A.: Metal additive manufacturing in aerospace: a review. Mater. Des. **209**, 110008 (2021). https://doi.org/10.1016/j.matdes.2021.110008
8. Maconachie, T., Leary, M., Lozanovski, B., Zhang, X., Qian, M., Faruque, O., Brandt, M.: SLM lattice structures: properties, performance, applications and challenges. Mater. Des. **183**, 108137 (2019). https://doi.org/10.1016/j.matdes.2019.108137
9. Mahmoud, D., Elbestawi, M.: Lattice structures and functionally graded materials applications in additive manufacturing of orthopedic implants: a review. J. Manuf. Mater. Process. **1**(2), 13 (2017). https://doi.org/10.3390/jmmp1020013
10. Pan, C., Han, Y., Lu, J.: Design and optimization of lattice structures: a review. Appl. Sci. **10**(18), 6374 (2020). https://doi.org/10.3390/app10186374
11. Maxwell, J.C.: L. On the calculation of the equilibrium and stiffness of frames. Lond. Edinb. Dublin Philos. Mag. J. Sci. **27**(182), 294–299 (1864). https://doi.org/10.1080/14786446408643668
12. Ashby, M.F.: The properties of foams and lattices. Philos. Trans. Ser. A, Math. Phys. Eng. Sci. **364**(1838), 15–30 (2006). https://doi.org/10.1098/rsta.2005.1678
13. Körner, C.: Additive manufacturing of metallic components by selective electron beam melting—a review. Int. Mater. Rev. **61**(5), 361–377 (2016). https://doi.org/10.1080/09506608.2016.1176289
14. List, F.A., Dehoff, R.R., Lowe, L.E., Sames, W.J.: Properties of Inconel 625 mesh structures grown by electron beam additive manufacturing. Mater. Sci. Eng., A **615**, 191–197 (2014). https://doi.org/10.1016/j.msea.2014.07.051
15. Cheng, X.Y., Li, S.J., Murr, L.E., Zhang, Z.B., Hao, Y.L., Yang, R., Medina, F., Wicker, R.B.: Compression deformation behavior of Ti-6Al-4V alloy with cellular structures fabricated by

electron beam melting. J. Mech. Behav. Biomed. Mater. **16**, 153–162 (2012). https://doi.org/10.1016/j.jmbbm.2012.10.005

16. Parthasarathy, J., Starly, B., Raman, S., Christensen, A.: Mechanical evaluation of porous titanium (Ti6Al4V) structures with electron beam melting (EBM). J. Mech. Behav. Biomed. Mater. **3**(3), 249–259 (2010). https://doi.org/10.1016/j.jmbbm.2009.10.006

17. Galarraga, H., Warren, R.J., Lados, D.A., Dehoff, R.R., Kirka, M.M., Nandwana, P.: Effects of heat treatments on microstructure and properties of Ti-6Al-4V ELI alloy fabricated by electron beam melting (EBM). Mater. Sci. Eng., A **685**, 417–428 (2017). https://doi.org/10.1016/j.msea.2017.01.019

18. Zhao, S., Li, S.J., Hou, W.T., Hao, Y.L., Yang, R., Misra, R.D.K.: The influence of cell morphology on the compressive fatigue behavior of Ti-6Al-4V meshes fabricated by electron beam melting. J. Mech. Behav. Biomed. Mater. **59**, 251–264 (2016). https://doi.org/10.1016/j.jmbbm.2016.01.034

19. Lietaert, K., Cutolo, A., Boey, D., van Hooreweder, B.: Fatigue life of additively manufactured Ti6Al4V scaffolds under tension-tension, tension-compression and compression-compression fatigue load. Sci. Rep. **8**(1), 4957 (2018). https://doi.org/10.1038/s41598-018-23414-2

20. Huynh, L., Rotella, J., Sangid, M.D.: Fatigue behavior of IN718 microtrusses produced via additive manufacturing. Mater. Des. **105**, 278–289 (2016). https://doi.org/10.1016/j.matdes.2016.05.032

21. Goodall, R., Hernandez-Nava, E., Jenkins, S.N.M., Sinclair, L., Tyrwhitt-Jones, E., Khodadadi, M.A., Ip, D.H., Ghadbeigi, H.: The effects of defects and damage in the mechanical behavior of Ti6Al4V lattices. Frontiers in Materials **6**, 66 (2019). https://doi.org/10.3389/fmats.2019.00117

22. Kotzem, D., Arold, T., Niendorf, T., Walther, F.: Damage tolerance evaluation of E-PBF-manufactured Inconel 718 strut geometries by advanced characterization techniques. Materials (Basel, Switzerland) **13**(1), 1–21 (2020). https://doi.org/10.3390/ma13010247

23. Kotzem, D., Ohlmeyer, H., Walther, F.: Damage tolerance evaluation of a unit cell plane based on electron beam powder bed fusion (E-PBF) manufactured Ti6Al4V alloy. Procedia Struct. Integr. **28**, 11–18 (2020). https://doi.org/10.1016/j.prostr.2020.10.003

24. Brockmann, S., Krupp, U. (eds.): Werkstoffprüfung 2021 - Werkstoffe und Bauteile auf dem Prüfstand. Stahlinstitut VDEh, Düsseldorf (2021)

25. Al-Juboori, L.A., Niendorf, T., Brenne, F.: On the tensile properties of Inconel 718 fabricated by EBM for as-built and heat-treated components. Metall. Mater. Trans. B. **49**(6), 2969–2974 (2018). https://doi.org/10.1007/s11663-018-1407-4

26. Helmer, H., Bauereiß, A., Singer, R.F., Körner, C.: Grain structure evolution in Inconel 718 during selective electron beam melting. Mater. Sci. Eng., A **668**, 180–187 (2016). https://doi.org/10.1016/j.msea.2016.05.046

27. Amato, K.N., Gaytan, S.M., Murr, L.E., Martinez, E., Shindo, P.W., Hernandez, J., Collins, S., Medina, F.: Microstructures and mechanical behavior of Inconel 718 fabricated by selective laser melting. Acta Mater. **60**(5), 2229–2239 (2012). https://doi.org/10.1016/j.actamat.2011.12.032

28. Sun, S.-H., Koizumi, Y., Saito, T., Yamanaka, K., Li, Y.-P., Cui, Y., Chiba, A.: Electron beam additive manufacturing of Inconel 718 alloy rods: Impact of build direction on microstructure and high-temperature tensile properties. Addit. Manuf. **23**, 457–470 (2018). https://doi.org/10.1016/j.addma.2018.08.017

29. Scala, F. (ed.): Fluidized Bed Technologies for Near-Zero Emission Combustion and Gasification. Woodhead Publishing Series in Energy, vol. 59. WP Woodhead Publ., Oxford (2013)

30. Hrabe, N., Quinn, T.: Effects of processing on microstructure and mechanical properties of a titanium alloy (Ti–6Al–4V) fabricated using electron beam melting (EBM), Part 2: Energy input, orientation, and location. Mater. Sci. Eng., A **573**, 271–277 (2013). https://doi.org/10.1016/j.msea.2013.02.065

31. Liu, S., Shin, Y.C.: Additive manufacturing of Ti6Al4V alloy: a review. Mater. Des. **164**, 107552 (2019). https://doi.org/10.1016/j.matdes.2018.107552

32. Polonsky, A.T., Echlin, M.P., Lenthe, W.C., Dehoff, R.R., Kirka, M.M., Pollock, T.M.: Defects and 3D structural inhomogeneity in electron beam additively manufactured Inconel 718. Mater. Charact. **143**, 171–181 (2018). https://doi.org/10.1016/j.matchar.2018.02.020

33. Persenot, T., Martin, G., Dendievel, R., Buffiére, J.-Y., Maire, E.: Enhancing the tensile properties of EBM as-built thin parts: effect of HIP and chemical etching. Mater. Charact. **143**, 82–93 (2018). https://doi.org/10.1016/j.matchar.2018.01.035

34. Kotzem, D., Dumke, P., Sepehri, P., Tenkamp, J., Walther, F.: Effect of miniaturization and surface roughness on the mechanical properties of the electron beam melted superalloy Inconel®718. Progr. Addit. Manuf. **117**, 371 (2019). https://doi.org/10.1007/s40964-019-001 01-w

35. Suard, M., Martin, G., Lhuissier, P., Dendievel, R., Vignat, F., Blandin, J.-J., Villeneuve, F.: Mechanical equivalent diameter of single struts for the stiffness prediction of lattice structures produced by electron beam melting. Addit. Manuf. **8**, 124–131 (2015). https://doi.org/10.1016/ j.addma.2015.10.002

36. Arabnejad, S., Burnett Johnston, R., Pura, J.A., Singh, B., Tanzer, M., Pasini, D.: High-strength porous biomaterials for bone replacement: a strategy to assess the interplay between cell morphology, mechanical properties, bone ingrowth and manufacturing constraints. Acta Biomater. **30**, 345–356 (2016). https://doi.org/10.1016/j.actbio.2015.10.048

37. Persenot, T., Burr, A., Martin, G., Buffiere, J.-Y., Dendievel, R., Maire, E.: Effect of build orientation on the fatigue properties of as-built electron beam melted Ti-6Al-4V alloy. Int. J. Fatigue **118**, 65–76 (2019). https://doi.org/10.1016/j.ijfatigue.2018.08.006

38. van Bael, S., Kerckhofs, G., Moesen, M., Pyka, G., Schrooten, J., Kruth, J.P.: Micro-CT-based improvement of geometrical and mechanical controllability of selective laser melted Ti6Al4V porous structures. Mater. Sci. Eng., A **528**(24), 7423–7431 (2011). https://doi.org/10.1016/j. msea.2011.06.045

39. Chahid, Y., Racasan, R., Pagani, L., Townsend, A., Liu, A., Bills, P., Blunt, L.: Parametrically designed surface topography on CAD models of additively manufactured lattice structures for improved design validation. Addit. Manuf. **37**, 101731 (2021). https://doi.org/10.1016/j. addma.2020.101731

40. Walther, F.: Microstructure-oriented fatigue assessment of construction materials and joints using short-time load increase procedure. Mater. Test. **56**, 519–527 (2014)

41. Piotrowski, A., Eifler, D.: Bewertung zyklischer Verformungsvorgänge metallischer Werkstoffe mit Hilfe mechanischer, thermometrischer und elektrischer meßverfahren Characterization of cyclic deformation behaviour by mechanical, thermometrical and electrical methods. Mat.-wiss. u. Werkstofftech. **26**, 121–127 (1995)

42. Basquin, O.H.: The exponential law of endurance tests. Proc. ASTM **10**, 625–630 (1910)

43. Walther, F., Eifler, D.: Cyclic deformation behavior of steels and light-metal alloys. Mater. Sci. Eng., A **468–470**, 259–266 (2007). https://doi.org/10.1016/j.msea.2006.06.146

44. Vendra, L., Neville, B., Rabiei, A.: Fatigue in aluminum–steel and steel–steel composite foams. Mater. Sci. Eng., A **517**(1–2), 146–153 (2009). https://doi.org/10.1016/j.msea.2009.03.075

45. Radlof, W., Panwitt, H., Benz, C., Sander, M.: Image-based and in-situ measurement techniques for the characterization of the damage behavior of additively manufactured lattice structures under fatigue loading. Procedia Struct. Integr. **38**, 50–59 (2022). https://doi.org/10.1016/j.pro str.2022.03.006

46. Yavari, S.A., Wauthle, R., van der Stok, J., Riemslag, A.C., Janssen, M., Mulier, M., Kruth, J.P., Schrooten, J., Weinans, H., Zadpoor, A.A.: Fatigue behavior of porous biomaterials manufactured using selective laser melting. Mater. Sci. Eng. C, Mater. Biol. Appl. **33**(8), 4849–4858 (2013). https://doi.org/10.1016/j.msec.2013.08.006

47. Hrabe, N.W., Heinl, P., Flinn, B., Körner, C., Bordia, R.K.: Compression-compression fatigue of selective electron beam melted cellular titanium (Ti-6Al-4V). J. Biomed. Mater. Res. Part B, Appl. Biomater. **99**(2), 313–320 (2011). https://doi.org/10.1002/jbm.b.31901

48. Sugimura, Y., Rabiei, A., Evans, A., Harte, A., Fleck, N.: Compression fatigue of a cellular Al alloy. Mater. Sci. Eng., A **269**(1–2), 38–48 (1999). https://doi.org/10.1016/S0921-5093(99)001 47-1

49. Benedetti, M., Du Plessis, A., Ritchie, R.O., Dallago, M., Razavi, S., Berto, F.: Architected cellular materials: a review on their mechanical properties towards fatigue-tolerant design and fabrication. Mater. Sci. Eng. R. Rep. **144**, 100606 (2021). https://doi.org/10.1016/j.mser.2021.100606

50. Dallago, M., Fontanari, V., Torresani, E., Leoni, M., Pederzolli, C., Potrich, C., Benedetti, M.: Fatigue and biological properties of Ti-6Al-4V ELI cellular structures with variously arranged cubic cells made by selective laser melting. J. Mech. Behav. Biomed. Mater. **78**, 381–394 (2018). https://doi.org/10.1016/j.jmbbm.2017.11.044

51. International Organization for Standardization: Mechanical testing of metals—Ductility testing—Compression test for porous and cellular metals (2011) (13314). Accessed 19 Mar 2020

52. Razavi, S., van Hooreweder, B., Berto, F.: Effect of build thickness and geometry on quasi-static and fatigue behavior of Ti-6Al-4V produced by electron beam melting. Addit. Manuf. **36**, 101426 (2020). https://doi.org/10.1016/j.addma.2020.101426

Enhanced Assessment of the Fatigue Behavior and Damage Tolerance of Additively Manufactured Metals and Components

M. Merghany, M. Teschke, F. Stern, J. Tenkamp, and F. Walther⬤

Abstract Layer-by-layer manufacturing of complex and lightweight structures is possible with additive manufacturing (AM) using powder bed fusion (PBF). Due to the fine microstructures generated by the high cooling rates, the tensile strength is improved, whereas the fatigue strength is comparable or even reduced. This is due to the presence of process-induced defects formulated during the manufacturing process in combination with the increased notch and mean stress sensitivity of high-strength metals. The damage evolution, including crack initiation and propagation, may be determined concerning fatigue stress and lifetime using modern measurement techniques before, during, and after fatigue testing. This will lead to a deeper understanding of the characterization of AM metals and how different variants and parameters can affect the fatigue behavior. Through this paper, three different AM alloys (AlSi10Mg, 316L and TNM-B1) are studied with respect to how the process-induced defects can affect the fatigue lifetime and resulting scattering and how the fatigue damage tolerance can be uniformly represented.

Keywords Additive manufacturing · AlSi10Mg · 316L · TNM-B1 · Fatigue behavior · Fatigue damage tolerance

---

M. Merghany · M. Teschke · F. Stern · J. Tenkamp · F. Walther (✉)
Chair of Materials Test Engineering, TU Dortmund University, Baroper Str. 303, 44227 Dortmund, Germany
e-mail: frank.walther@tu-dortmund.de

M. Merghany
e-mail: mohamed.merghany@tu-dortmund.de

M. Teschke
e-mail: mirko.teschke@tu-dortmund.de

F. Stern
e-mail: felix.stern@tu-dortmund.de

J. Tenkamp
e-mail: jochen.tenkamp@tu-dortmund.de

1 Introduction

The additive manufacturing (AM) process is building the components layer-by-layer. Therefore, it can produce complex near-net-shapes (e.g. internal cooling or heating channels, bionic structures, dental implants) that the conventional manufacturing processes cannot do [1, 2]. AM is now growing very rapidly in terms of produced parts and economic importance [3]. Laser powder bed fusion of metals (PBF-LB) and electron beam powder bed fusion (PBF-EB) are competitive processes due to their ability to produce complex and precious parts. While PBF-LB shows better surface roughness and lower production costs compared to PBF-EB [4], the high process temperatures and vacuum atmosphere prevent cracking and oxidation in PBF-EB [5].

Although AM enables the fabrication of arbitrary 3D-structures with unprecedented degrees of freedom, this can lead to structure and design issues inconclusively answered with respect to part density, pores distribution and fatigue behavior [6]. In addition to the pore size, pore shape, and pore distribution, the distance between the pores and the surface also plays a crucial role that influences the fatigue behavior. The effect of defects on the fatigue strength (σ_{aL}^*) can be calculated with the approach of Murakami et al. [7] and the extension for lightweight alloys according to Noguchi et al. [8] as described in Eq. 1.

$$\sigma_{aL}^* = \frac{C \cdot (HV + 120 \cdot E/E_{St})}{(a_i)^{1/6}} \quad \begin{array}{l} \text{Surface defect: } C = 1.43 \\ \text{Volume defect: } C = 1.56 \end{array} \tag{1}$$

where (HV) is Vickers hardness, (E) is the Young's modulus, (E_{St}) is Young's modulus of steel equal to 206 GPa. While the square-root of the projected cross-sectional area (A_i) of the fracture-inducing defect perpendicular to the loading direction equals to the equivalent defect size ($a_i = \sqrt{A_i}$).

Moreover, the model introduced by Shiozawa [9] for crack propagation-dominant parts can be used for a better representation and interpretation of the fatigue results. Here, as shown in Eq. 2 using various assumptions, the Paris-Erdogan law for describing the crack propagation behavior is integrated from the initial defect size to the critical crack size.

$$\Delta K_i = \left[\frac{C(m-2)}{2} \right]^{-1/m} \cdot \left(\frac{N_f}{a_i} \right)^{-1/m} \tag{2}$$

where (ΔK_i) is the cyclic stress intensity factor (SIF), (C) and (m) are assumptions for Paris law coefficients, and (N_f) is the number of cycles to failure. While (N_f/a_i) is the so-called defect-related fatigue life. The Paris law coefficients can be obtained via power-law by plotting SIF (ΔK_i) on Y-axis and (N_f/a_i) on X-axis using a log–log scale. Furthermore, SIF (ΔK_i) can be determined using Eq. 3.

$$\Delta K_i = \Delta \sigma \cdot \sqrt{\pi \cdot a_i} \cdot Y \quad \begin{array}{l} \text{Surface defect: } Y = 0.65 \\ \text{Volume defect: } Y = 0.50 \end{array} \tag{3}$$

In summary, different parameters like process parameters, building orientation and defect size influence the fatigue behavior of AM parts. Through this research, three alloys (AlSi10Mg, 316L and TNM-B1) were studied by applying the Shiozawa and Murakami-Noguchi models and investigating the effect of critical defect size on applying these models. In addition to the fatigue testing, fractographic analyses using scanning electron microscopy were carried out in order to identify the damage mechanisms with regard to fatigue crack initiation and propagation and to define the size and type of the defect led to failure.

2 Materials and Methods

2.1 Materials

Three additively manufactured alloys are investigated: the aluminum alloy AlSi10Mg, the stainless steel AISI 316L and the titanium aluminide alloy TNM-B1 (Ti-46.5Al-4Nb-1Mo-0.1B). AlSi10Mg specimens were produced with PBF-LB using EOS M290 system. 316L specimens were also manufactured by PBF-LB using M2 Cusing system at three different orientations to the building direction (0°, 45°, and 90°). Finally, TNM-B1 specimens were manufactured by PBF-EB on an Arcam A2X system.

AlSi10Mg, 316L, and half of TNM-B1 specimens were investigated in the as-built condition without post-process heat treatment, while half of the TNM-B1 specimens were hot isostatically pressed (HIP). All specimens were tested at room temperature (RT).

2.2 Methods

2.2.1 Fatigue Testing

For AlSi10Mg alloy, stress-controlled fatigue tests were performed using a Rumul Testronic 150 kN resonance testing system with 20 kN load cell at a test frequency of $f = 70$ Hz and stress ratio of $R = -1$ (fully-reversed loading) up to 10^7 cycles.

For 316L steel, stress-controlled fatigue tests were carried out on the servo-hydraulic testing system Schenck PSB100 with Instron 8800 controller (force ± 75 kN) at a test frequency of $f = 10$ Hz and stress ratio of $R = -1$ up to $3 \cdot 10^6$ cycles.

For TNM-B1 alloy, the fatigue tests were performed on the servo-hydraulic testing system Instron 8801 (force ± 100 kN). Stress-controlled fatigue tests were performed at a test frequency of $f = 20$ Hz and the stress ratio of $R = -1$ up to $2 \cdot 10^6$ cycles.

2.2.2 Macro-hardness Measurement

The hardness of the three alloys was determined by macro-hardness HV10 measurements using the Wolpert Dia-Testor 2Rc Vickers hardness testing system. For each alloy, the hardness measurements were performed according to DIN EN ISO 6507-1, then the mean values and standard of deviations were calculated.

2.2.3 Fractographic Analysis

The microstructure and fracture surfaces of all specimens were examined using scanning electron microscope (SEM) Tescan MIRA3 XMU to determine the location, size, and shape of the defect which led to failure. These parameters were determined by post-processing of the obtained SEM images using ImageJ software. The porosity and defect distributions were characterized using microfocus-computed tomography (μ-CT) Nikon XT H160, where the parts' relative densities and the equivalent defect size a_i were also quantified.

3   Results and Discussion

3.1   Hardness and Fractography Analysis Results

The results of Vickers hardness measurements for the three alloys and the three orientations for 316L ($0°$, $45°$ and $90°$) are shown in Fig. 1a. It is observed that TNM-B1 showed the highest hardness compared to the other alloys equals to 418 $\pm$ 6 HV10 for the AB condition and 385 $\pm$ 4 HV10 for the HIP condition. This concludes that HIP decreases the hardness. AlSi10Mg has the lowest hardness value equals to 106 $\pm$ 3 HV10. While 316L has an intermediate hardness value equals to 211 $\pm$ 5 HV10 but it is obtained that the hardness is slightly decreasing with increasing building direction orientation angles from $0°$ to $90°$.

Furthermore, the fractographic analysis results represented by the average size of the fracture-inducing defect are shown in Fig. 1b. It is observed that the crack initiating defect sizes for TNM-B1 are the lowest with 64 $\pm$ 25 μm for the AB condition and 20 $\pm$ 8 for the HIP condition. Moreover, AlSi10Mg has the highest defect size equals to 242 $\pm$ 133 μm. Additionally, the highest variation in the defect sizes is observed in AlSi10Mg and 316L which will cause scattering in the fatigue results, as will be shown in the next section.

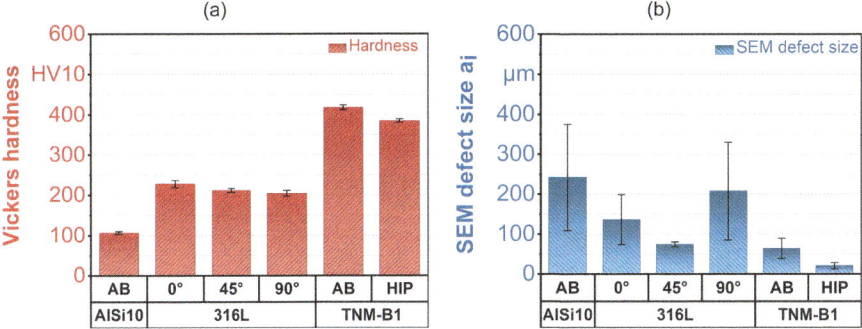

Fig. 1 **a** Vickers hardness and **b** fractographic analysis for AlSi10Mg, 316L, TNM-B1

3.2 Fatigue Testing Results

After fatigue testing of the three alloys at different stress amplitudes with stress ratio $R = -1$, Woehler curves are constructed by plotting stress amplitude σ_a versus the number of cycles to failure N_f on log–log scale. It is obtained from Fig. 2a that AlSi10Mg has the lowest fatigue strength due to the low hardness and large defect size. The verifying defect size was responsible for increasing the scattering within the Woehler curve. While in Fig. 2c, TNM-B1 has the highest fatigue strength and lifetime compared to 316L and AlSi10Mg and this correlates with its higher hardness and smaller defect size. By reducing the number and size of the defects with HIP, the fatigue strength at $(2 \cdot 10^6)$ could be further increased by 42% to 500 MPa. It is shown in Fig. 2b, that the present defects cause high scattering (coefficient of determination $R^2 = 0.09$) in the fatigue results of 316L and this is mainly because of the high variation in the defect sizes as shown in Fig. 1b. The main reason for this is the presence of flat defects perpendicular to the build direction, which then are differently orientated in the later fatigue tests. By that, 90° specimens show the worst fatigue lifetime as the defects are larger because of their orientation compared to the load direction.

It is concluded from the previous graphs that a uniform fatigue assessment based on classical Woehler curves is not possible because the high scattering in the fatigue results generated from different defect sizes and orientations of those defects for the 316L which led to failure.

3.3 Murakami-Noguchi and Shiozawa Approaches

For a more uniform fatigue assessment, relative Woehler curve and Shiozawa diagram are constructed as shown in Fig. 3. The relative Woehler curve in Fig. 3a is generated by dividing the Y-axis of the Woehler curve by the fatigue limit calculated using

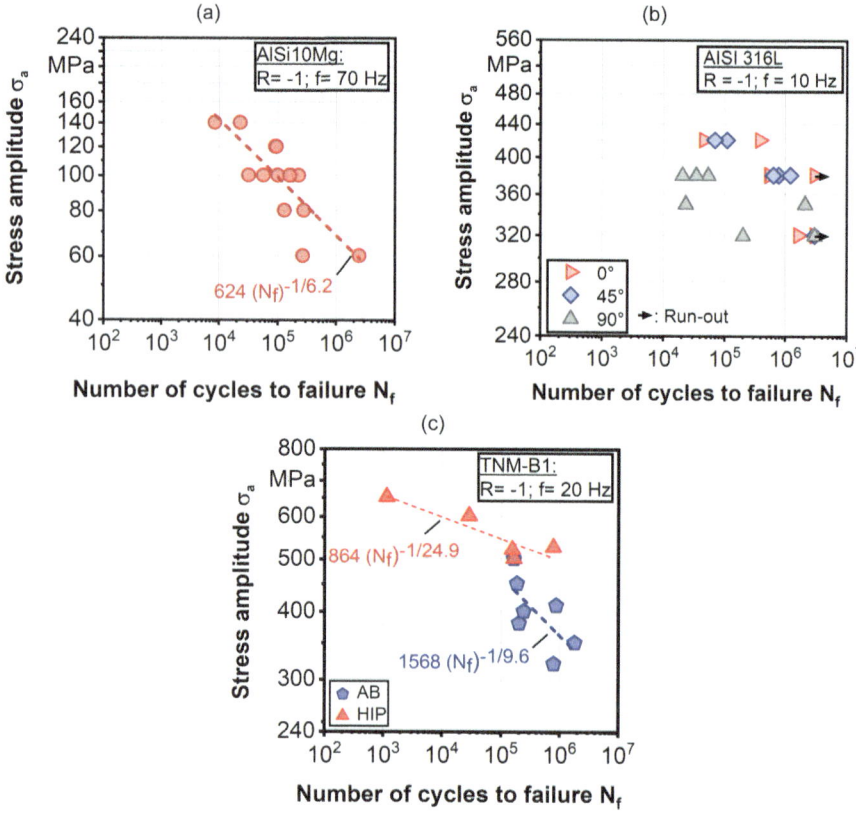

Fig. 2 Woehler curves for **a** AlSi10Mg, **b** 316L [10, 11], **c** TNM-B1 [12]

Eq. 1. According to the relative Woehler curve, 316L showed overestimated fatigue damage tolerance as the calculated fatigue strength normally belongs to 10^7 cycles, while, AlSi10Mg and TNM-B1 showed underestimated fatigue damage tolerance. Moreover, the relative Woehler curve based on Murakami-Noguchi approach cannot describe the run-out specimens as at maximum number of cycles equal to 10^7, the relative stress amplitude should be equal to 1.0 and the model has to be modified to overcome this point. In summary, the scattering is significantly reduced using Murakami-Noguchi model, but the approach cannot be used to directly compare different Fe-, Al- and Ti-alloys. However, it could still be that the Al- and Ti-alloy fatigue life is much more dominated by defect size.

Therefore, using Shiozawa curves which also consider the failure-initiating defect size can describe the damage tolerance more efficiently for both alloys AlSi10Mg and 316L as shown in Fig. 3b. While in TNM-B1, the effect of defects is very large but by HIP defect size is reduced and the fatigue behavior is improved by slightly decreasing the hardness. Shiozawa curve is corrected for the defect size, therefore, it was able to decrease the scattering in the Woehler curve that was generated from

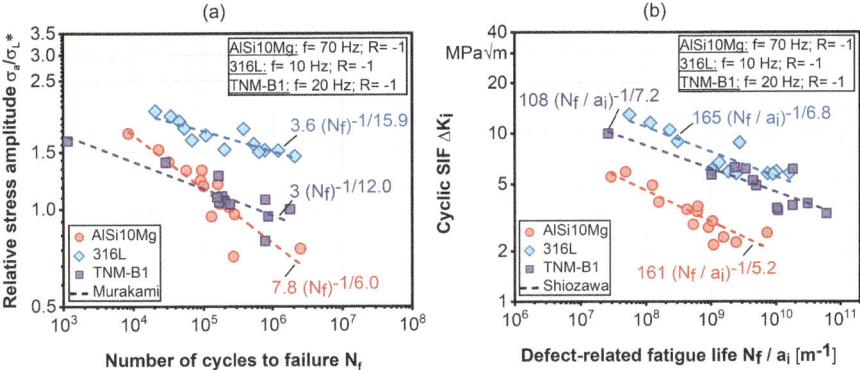

Fig. 3 Uniform evaluation of damage tolerance for AlSi10Mg, 316L and TNM-B1 [12]: **a** relative Woehler curves based on Murakami-Noguchi, **b** Shiozawa curve

the different defect sizes. This can be observed especially in 316L alloy where the coefficient of determination improved from $R^2 = 0.09$ in Woehler curve into $R^2 = 0.78$ in the Shiozawa curve. It was also obtained that the Shiozawa curve can uniformly plot the damage tolerance of the three different alloys by decreasing the scattering in the fatigue data resulting from the different defect sizes.

4 Conclusion and Outlook

For summary, it was found that the effect of defects in AM alloys has a high influence on the fatigue lifetime. It is also obtained that the building orientation also affects the orientation of the internal critical defects that may lead to fatigue failure. Moreover, the alloys that have larger defect sizes have more scattering in the fatigue results of the Woehler curves. As a result, the defect size a_i introduced by Murakami is particularly suitable for the geometric description of the defect as an initial defect or crack size while the models presented by both Shiozawa and Murakami take this influence into account and enable a "defect-based" representation of the fatigue behavior.

In future investigations, the influence of the relative part density and alloy specific density on fatigue behavior should be investigated.

Acknowledgements The authors thank the German Research Foundation (Deutsche Forschungs-gemeinschaft, DFG) for its financial support within the research projects "Mechanism-based investigation of additively-manufactured aluminum matrix composites (AMC) for enhanced mechanical strength" (project no. 425479688), "Mechanism-based assessment of the influence of powder production and process parameters on the microstructure and the deformation behavior of SLM-compacted C + N steels in air and in corrosive environments" (project no. 372290567) and "Microstructure and defect controlled additive manufacturing of gamma titanium aluminides for function-based control of local materials properties" (project no. 404665753).

References

1. Wang, X., Xu, S., Zhou, S., Xu, W., Leary, M., Choong, P., Qian, M., Brandt, M., Xie, Y.M.: Topological design and additive manufacturing of porous metals for bone scaffolds and orthopaedic implants: a review. Biomaterials **83**, 127–141 (2016)
2. Bose, S., Vahabzadeh, S., Bandyopadhyay, A.: Bone tissue engineering using 3D printing. Mater. Today **16**(12), 496–504 (2013)
3. Bassoli, E., Denti, L., Comin, A., Sola, A., Tognoli, E.: Fatigue behavior of as-built L-PBF A357.0 parts. Metals **8**, 634 (2018)
4. Merkt, S., Hinke, C., Bültmann, J., Brandt, M., Xie, Y.M.: Mechanical response of TiAl6V4 lattice structures manufactured by selective laser melting in quasistatic and dynamic compression tests. J. Laser Appl. **27**, 1–7 (2015)
5. Körner, C.: Additive manufacturing of metallic components by selective electron beam melting—a review. Int. Mater. Rev. **61**, 361–377 (2016)
6. Aboulkhair, N.T., Simonelli, M., Parry, L., Ashcroft, I., Tuck, C., Hague, R.: 3D printing of aluminium alloys: additive manufacturing of aluminium alloys using selective laser melting. Prog. Mater Sci. **106**, 100578 (2019)
7. Murakami, Y.: Material defects as the basis of fatigue design. Int. J. Fatigue **41**, 2–10 (2012)
8. Noguchi, H., Morishige, K., Fujii, T., Kawazoe, T., Hamada, S.: Proposal of method for estimation stress intensity factor range on small crack for light metals. In: Proceedings of the 56th JSMS Annual Meetings, pp. 137–138 (2007)
9. Shiozawa, K., Lu, L.: Effect of non-metallic inclusion size and residual stresses on giga-cycle fatigue properties in high strength steel. In: 11th International Fatigue Congress, pp. 44–46, 33–42 (2008)
10. Stern, F., Kleinhorst, J., Tenkamp, J., Walther, F.: Investigation of the anisotropic cyclic damage behavior of selective laser melted AISI 316L stainless steel. Fatigue Fract. Eng. Mater. Struct. **42**, 2422–2430 (2019)
11. Stern, F., Tenkamp, J., Walther, F.: Non-destructive characterization of process-induced defects and their effect on the fatigue behavior of austenitic steel 316L made by laser-powder bed fusion. Progr. Addit. Manuf. **5**, 287–294 (2020)
12. Teschke, M., Moritz, J., Tenkamp, J., Marquardt, A., Leyens, C., Walther, F.: Defect-based characterization of the fatigue behavior of additively manufactured titanium aluminides. Int. J. Fatigue, **107047**, 1–9 (2022)

Effects of Printing Parameters on the Quality of FFF Printed Parts with Red PLA Filaments from Different Suppliers

João Pedro Ramalho, Leonardo Santana⊙, Henrique Takashi Idogava⊙, and Jorge Lino Alves⊙

Abstract The development of open-source movements in Fused Filament Fabrication (FFF) was important for the popularization of the technology for different levels of users, but also for motivating the emergence of companies focused on manufacturing machines and printing materials. Many material options are available on the market, with varied polymer bases and types of additives, among which pigments stand out. However, frequently, the effect of the pigment in the parameterization process and in the final quality of parts is neglected by the suppliers, and that often becomes a problem for printing. Recent studies show that for the same supplier, color variation in Polylactic acid (PLA) filaments is a significant factor for changes in mechanical properties, especially when the pigment is red. There was an opportunity to develop a study in which the color factor is fixed (red), and the influence of different filament suppliers is evaluated in the parametric configuration planning and, consequently, in the integrity of the parts obtained. Red PLA filaments of three brands were selected, which were input elements for a three-part investigation. First, the filaments were characterized by Fourier Transform Infrared Spectroscopy (FTIR), Dynamic Mechanical Analysis (DMA), and density measurement. In the second moment, a Taguchi experiment was used to evaluate the flexural mechanical properties and mass of the parts, considering the variation of layer thickness, extrusion temperature, print speed and extrusion multiplier, each one with three levels. The best levels of phase two parameter for each supplier were used to create a verification print model which manufactured parts were evaluated by the same responses

---

J. P. Ramalho · J. L. Alves (✉)
Faculdade de Engenharia da Universidade do Porto (FEUP), 4200-465 Porto, Portugal
e-mail: falves@fe.up.pt

J. P. Ramalho
e-mail: up201303583@edu.fe.up.pt

L. Santana · H. T. Idogava
Escola de Engenharia de São Calos da Universidade de São Paulo (USP), 13566-590 São Carlos (SP), Brasil

L. Santana · J. L. Alves
Laboratório Associado em Energia, Transportes e Aeronáutica, INEGI, 4200-465 Porto, Portugal

L. F. M. da Silva et al. (eds.), *1st International Conference on Engineering Manufacture 2022*, Proceedings in Engineering Mechanics,
https://doi.org/10.1007/978-3-031-13234-6_3

as in the previous step. Finally, the results were compared to those from a fixed parameter setting from the literature. FTIR, DMA, and density analysis showed that red PLAs from different suppliers have similar responses. Parametrically, the red PLA filaments behaved differently, requiring, in some cases, specifics machines to produce the parts. In all cases, the verification models made the responses better than those identified with a fixed setting from the literature, which demonstrates the need for the user to adapt the system according to the supplier chosen, especially in controlling the extrusion flow rate. In summary, despite using the same polymer base (PLA), different suppliers possibly use their own sets of additives either to modify the color or stabilize the material. This affects the rheology and thermal behavior of the polymers, which requires unique printing settings, altering mechanical properties, and mass of parts. The study also highlights the importance of considering the determination of quality regulatory mechanisms in the production and commercialization of pigmented thermoplastic filaments for 3D FFF printing.

Keywords FFF · PLA · Filament supplier · Pigment · Parametric setting

1 Introduction

Additive manufacturing, also known as 3D printing, is the combination of techniques that add layers of material to build complex parts. Among the additive processes, the FFF (Fused Filament Fabrication) stands out, in which the material in the form of a filament is heated, extruded, and deposited under a printing bed using a coordinate system. This process is considered the most used due to open-source movements that reduced the price of equipment and popularized its use. Open-source movements have also allowed varying printing materials and seeking new functions for printed parts. The start point of 3D printing is the virtual model, which inserts the process into a computational environment. This environment is conducive for online printing platforms to share digital files and encourage home users [1].

Following the same line of popularization of 3D printers, it is also possible to observe the development of new slicing software and filaments with different properties [2]. Algorithms that allow mixing colors in different regions of the object show new opportunities for using color palettes that are controlled by path parameters [3, 4]. However, the impulse of companies to manufacture new materials without appropriate tests results in difficulties for the user and loss of quality properties of the parts. There is a critical printing temperature for each color and according to Wittbrodt and Pearce [5], the lack of data and standards on commercial printers limit the studies of more sophisticated designs and equipment.

Many studies address the printing parameters that can be changed during the slicing process of virtual models. But few comment on the relationship between material properties and parameters, such as the influence of filament color. Davis et al. [6] relates the formulation of different printing filaments with volatile organic compound emissions. It was observed that even for the same material there can be a great difference in composition for different brands.

Following the same line, Hanon et al. [7] investigated the accuracy of specimens manufactured with variations of printing parameters in gray, white and black PLA. The conclusion of the study points the black filament as the material with the smallest variations and that the dimensional variation of parts printed of different colors can be minimized with the correct setup of the material parameters.

The characterization of non-processed filaments in blue, gray, transparent, orange, and natural colors was the object of study by Matos et al. [8]. Although the properties of the thermoplastic material are changed in the extrusion process, it is interesting to note different levels of crystallinity and glass transition temperatures of PLAs.

Two interesting studies have correlated characteristics associated with commercial filaments for FFF, an aspect that directly affects the process behavior and the user's relationship with the 3D printer. Bermudez et al. [9] evaluated mechanical, rheological, chemical properties, and crystallinity levels of two grades of Polylactic acid (PLA), 4043D and 3D870. They showed that the different grades differ in mechanical properties, melt flow index, and crystallinity level, among others. Castro [10], on the other hand, evaluated PLA filaments from the same supplier, varying their colors: white, blue, red, and orange. The color significantly influenced the flexural mechanical properties of parts after 3D printing and after heat treatments. Castro [10] also warns that red is the color that induces the greatest challenges to the processing of filaments.

Despite advances in filament characterization studies, especially in terms of color, few studies systematically assess the influence of the material brand, that is, what the user has available to buy in the market related to FFF technology. There is a need to highlight the particularities of each material, even with the same polymeric base, when changing colors or suppliers. Adjustments in process parameters and 3D printing equipment favor better results in terms of quality of the parts obtained.

Based on this gap, this study proposes the evaluation of PLA filaments from different suppliers through the characterization of materials and analysis of the responses of parts printed as a function of the variation of process parameters. To show the influence of the material manufacturers, the study fixed the color factor in red, that is, the most critical factor described in the literature.

2 Materials and Methods

Three red Polylactic acid (PLA) filaments (1.75 mm diameter) from three different suppliers were evaluated: *Filament PM* ™ (PM),[1] *Kexcelled*™ (KE) and *Fill3D*™

---

[1] PM, KE and F3D: codes assigned by the authors to facilitate the process of referencing materials throughout the discussions of this paper.

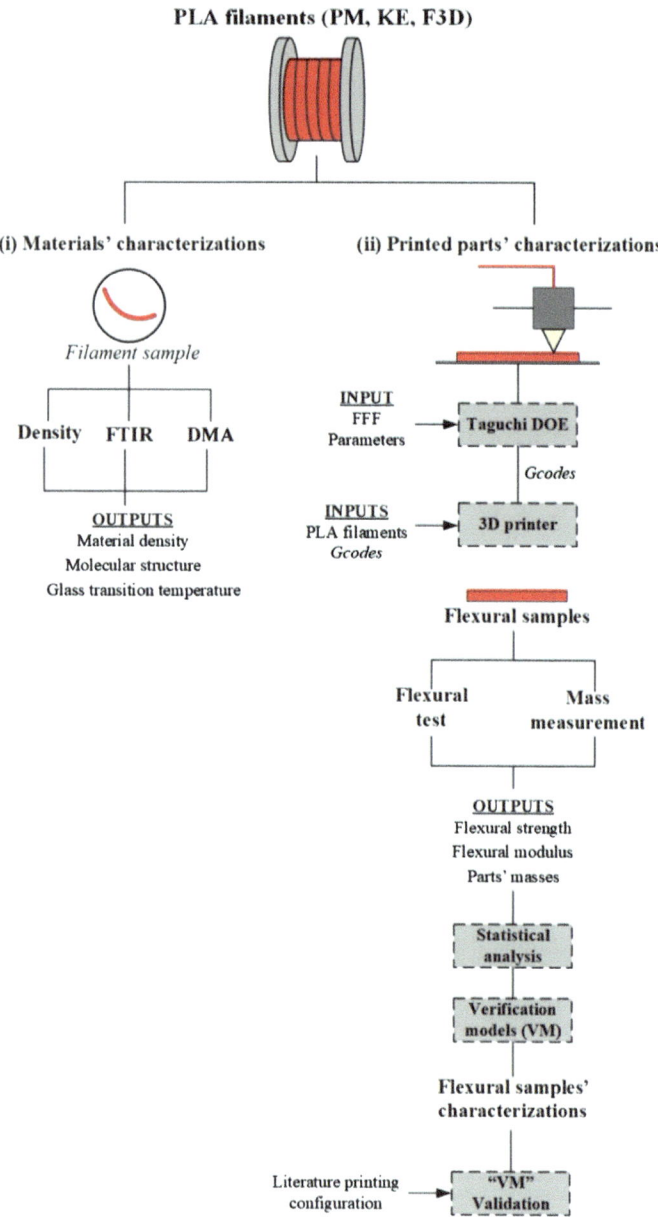

Fig. 1 Research methodology steps

(*Tucab*) (F3D). These filaments were applied in two stages of analysis: (i) character-ization of materials and (ii) evaluation of the quality of printed parts as a function of the variation of FFF process parameters (Fig. 1).

The three PLA filaments were characterized by density measurement, molecular structure analysis by Fourier Transform Infrared Spectroscopy (FTIR) and the glass transition temperature (Tg) determined by Dynamic Mechanical Analysis (DMA). The density measurement was performed using a XS205 *dual range* Mettler Toledo™ weighing scale (d = 0.01 mg/0.1 mg) with a dedicated setup for applying the Archimedes principle. For each filament, five density values were obtained (samples with 3 cm length), using distilled water (22 °C) as standard liquid for the test.

In the FTIR analysis, spectra were evaluated in a wavelength range from 4000 to 650 cm^{-1} in an Agilent Technologies™ Cary 630 equipment. For the DMA tests, samples were printed with parallelepiped geometry with dimensions of (35 × 5 × 1.3) mm. The test was conducted on a Netzsch™ DMA 242 E Artemis equipment in tensile mode. The conditions of the DMA test were temperature range from 25 to 90 °C (heating rate of 3 K/min), amplitude of 60 μm, and frequency of 1 Hz.

After individual characterization, the filaments were evaluated when building parts. The purpose of this experiment was to verify if the same polymeric base, in this case PLA, would require different levels of configuration of the 3D printing parameters due to the use of different material suppliers.

For this, samples were built for flexural tests, with geometries according to ASTM D790 [11], varying: extrusion temperature (Et), extrusion multiplier (Em), printing speed (Ps) and layer thickness (Lt). For each of the four factors, three levels of adjustment were evaluated. The combination between factors and their levels was performed with a L9 Taguchi orthogonal array (Table 1). For each of the nine conditions, three samples were manufactured simultaneously (Fig. 2), totalizing 27 parts for each of the three PLA filaments evaluated.

Table 1 Factor, levels, and L9 Taguchi orthogonal array

Factor	Level (1)	Level (2)	Level (3)
Et (°C)	215	225	235
Lt (mm)	0.1	0.2	0.3
Ps (mm/s)	40	60	80
Em (d.u.)	0.9	1.0	1.1

L9 Taguchi orthogonal array

Condition	Et (°C)	Lt (mm)	Ps (mm/s)	Em (d.u.)
1	215	0.1	40	0.9
2	215	0.2	60	1
3	215	0.3	80	1.1
4	225	0.1	40	1.1
5	225	0.2	60	0.9
6	225	0.3	80	1
7	235	0.1	40	1
8	235	0.2	60	1.1
9	235	0.3	80	0.9

Fig. 2 Set of three samples
printed for each condition of
the Taguchi array

The other printing variables were kept fixed according to the best results for PLA stated by Santana et al. [12–14] and Castro [10], in particular: number of perimeters (3), build orientation (lateral), infill pattern (concentric) and heated bed temperature (60 °C). The process planning of the models to be printed in each condition of Table 1 was performed in the PrusaSlicer software (version 2.4.1). An I3 MK3s Prusa 3D printer was used to manufacture samples with PM and KE filaments. In the case of the F3D filament, due to fluidity problems, material swelling and nozzle clogging, the test parts were produced on an I3 MK3s Prusa machine with a high-performance Mosquito™ hotend.

The effects of the parameters studied in this paper, Table 1, were evaluated as a function of responses such as the flexural modulus (FM), maximum flexural strength (MFS), and mass of the parts. The mass measurement of the parts was performed on a Scale House HLD300 weighing scale (resolution of ±0.05 g). The flexural tests were conducted in three-point mode on a Mecmesin™ test bench with a dynamometer with a load cell of 2.5 kN, with a 5 mm/min speed.

The results of mechanical testing and mass analysis were used to develop a verification model (VM), or global print configuration, for each of the analyzed filaments (Table 2). For this, the values of FM, MFS and mass were statistically analyzed with the aim: (i) to verify if the factors of the Taguchi experiment were significant for the variation of the responses, and (ii) to identify the levels of significant factors that improved the response performance. The highest number of recurrences among the best levels of each parameter, in each response, was the main criterion for selecting the adjustment of the impression variable for the VM.

The verification models in the Table 2 were used to print sets of six samples for flexural tests and mass measurement. To validate the VM of each filament, the results of the analysis were compared to those achieved with the 3D printing of samples with a fixed parameterization in the literature. In this sense, test parts were built with PM, KE and F3D, based on the printing configuration defined in the Castro [10]

Table 2 Verification model for each evaluated PLA filament supplier

Factor	Filament supplier		
	PM	KE	F3D
Et (°C)	235	235	225
Lt (mm)	0.1	0.1	0.1
Ps (mm/s)	60	60	60
Em (d.u.)	1	1.1	1.1

study[2]: number of perimeters (3), perimeters speed (20 mm/s), infill speed (40 mm/s), extrusion temperature (215 °C/210 °C), bed temperature (60 °C), extrusion multiplier (1), infill pattern (concentric), fill density (100%), build orientation (lateral), and layer thickness (0.2 mm). Six samples of each material with fixed configuration were also printed.

3 Results and Discussions

3.1 Filament Characterization

Through the FTIR analysis, Fig. 3, it was verified that the three filaments evaluated in this study presented the main molecules that compose the polymeric structure of PLA (Table 3): –CH stretching (CH_3 group) (*Zone I*), carbonyl stretching C=O (*Zone II*), CH_3 bending vibrations (*Zone III*), C–O–C stretching of the ester group (*Zone IV*). Another important spectroscopy response was the identification of the three thermoplastics as semi-crystalline polymers, a characteristic evidenced by *the Zone V* bands, which correspond to the amorphous (≈ 861 cm^{-1}, ≈ 863 cm^{-1}) and crystalline (≈ 756 cm^{-1}, ≈ 758 cm^{-1}) regions of PLA [10, 12, 15–18].

The red pigments used by the different suppliers do not significantly change the molecular structure of PLA. This finding is supported by the molecular similarity of these materials with those of other colored PLA filaments evaluated in the literature, such as the examples by Santana et al. [12] (blue) and Castro [10] (blue, white, orange and red). Furthermore, when comparing the three red filaments, the occurrence of one or more specific molecules that differentiated them from each other was not observed. In the case of the red PM filament, there is still a good molecular repeatability between the batches used in this study and in the work by Castro [10].

Perhaps the greatest influence of the pigments of each supplier is in the variation of the intensity and shades of the red color. This behavior can be seen by the difference in the size of the peaks (Fig. 3) for the same spectrum in each PLA. Even the variation in the intensity of the peaks occurs for the same supplier, as in the case of the two

---

[2] It is important to mention that the Castro [10] study printing configuration was selected because the author also used the red PLA filament from the Filament PM™ brand in his research.

Fig. 3 FTIR analysis curves
for PM, KE, F3D filaments
and for PLA characterized
by Castro [10]

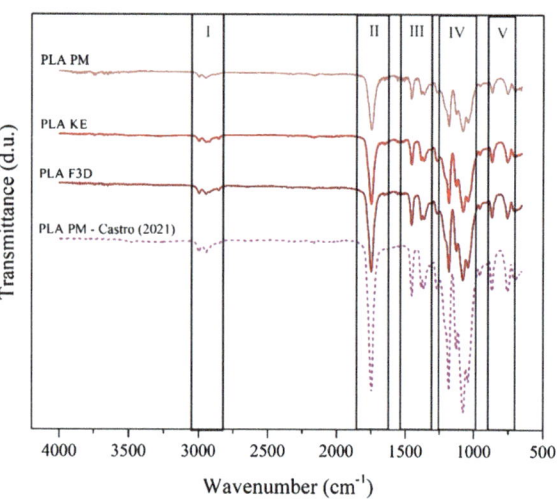

Table 3 Synthesis of FTIR analysis bands by spectrum regions

Filament supplier	Wavenumber (cm^{-1})				
	Zone I	Zone II	Zone III	Zone IV	Zone V
PM	2998; 2949	1749	1456; 1383; 1363	1183; 1126; 1085; 1042	867; 757
KE	2998; 2947	1744	1452; 1381; 1363	1180; 1128; 1083; 1040	867; 753
F3D	2998; 2947	1747	1456; 1381; 1363	1180; 1128; 1081; 1042	867; 755
PM Castro [10]	2998; 2947	1745	1452; 1381; 1359	1180; 1128, 1079, 1042	867; 755

evaluated PMs. The last scenario could indicate the lack of standardization in the use of pigments or variations in their concentrations.

The graph in Fig. 4 presents the average values of the density measurement of the red PLA filaments of the different suppliers.

The density of the filaments (Fig. 4), in turn, was close to the values provided by the material suppliers in their datasheets: 1.24 g/cm^3 (PM) [19], 1.23 a 1.26 g/cm^3 (KE) [20] and 1.24 g/cm^3 (F3D) [21]. In addition, even with the presence of pigments, the results of measuring the density of the filaments do not differ from the value described in the literature for PLA, that is, 1.25 g/cm^3 [22]. However, a greater dispersion in the density values of the F3D material is observed. This effect might be related to a non-uniform crystallization process, either by the processing conditions (irregular cooling, for example), by the presence of impurities, or by the influence of the additives used to formulate the material.

Fig. 4 Results of density measurement of red PLA filaments

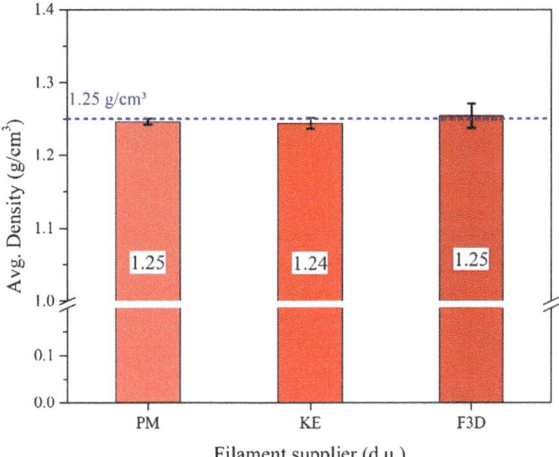

Finally, the DMA analysis shows that the PM, KE, and F3D PLAs presented glass transition temperature (Tg) values—obtained at the peak of tan(δ)—very close, being, respectively, 75.51, 75.65 and 75.69 °C. This characteristic could be related to the fact that Tg was measured in printed parts. As the polymers were processed under similar conditions (extrusion temperature 215/210 °C and base 60 °C), the new thermal history introduced by 3D Printing might have homogenized the temperatures. However, it should be noted that the red PLAs present glass transitions superior to those analyzed in the literature for printed parts with virgin PLA; in these cases, values of 62 °C [23] or intervals between 64 and 68 °C [24].

3.2 Parametric Study

3.2.1 PM Filament

The values of flexural modulus (FM), maximum flexural strength (MFS) and mass of samples in Taguchi's experiment were statistically evaluated by analysis of variance (ANOVA), as shown in Table 4. The graphs in Fig. 5 show the average values of FM (a), MFS (b) and mass by significant factors and their levels. The letters in the bars of the graph correspond to Tukey's post-hoc test. In the same factor, bars with the same letters are equivalent to same averages.

According to the performed ANOVA, Table 4, all FFF parameters evaluated were significant (P value < 0.05) for variations in mechanical properties of printed parts. Layer thickness (Lt) was the factor with the highest percentage of contribution to the modulus (P (%) = 87.03) and the maximum flexural stress (P (%) = 81.58).

The lower the value of "Lt", the higher the values of FM and MFS, Fig. 5a, b. By adjusting a small value for the layer thickness, the user is increasing the number

Table 4 ANOVA for FM, MFS and mass, PM filament responses

Factor	ANOVA ($\alpha = 95\%$) for FM					
	Df	SS	V	F	P	P (%)
Et	2	0.10	0.05	27.65	0.00	4.13
Lt	2	2.23	1.12	582.76	0.00	**87.03**
Ps	2	0.11	0.05	27.81	0.00	4.15
Em	2	0.09	0.04	22.39	0.00	3.34
Error	18	0.03	0.002			1.34
Total	26	2.56				100
Factor	ANOVA ($\alpha = 95\%$) for MFS					
Et	2	102.38	51.19	56.98	0.00	5.96
Lt	2	1402.13	701.07	780.32	0.00	**81.58**
Ps	2	34.31	17.15	19.09	0.00	2.00
Em	2	163.76	81.88	91.14	0.00	**9.53**
Error	18	16.17	0.90			0.94
Total	26	1718.75				100
Factor	ANOVA ($\alpha = 95\%$) for Mass					
Et	2	0.02	0.01	274.20	0.00	0.28
Lt	2	1.11	0.55	17,111.40	0.00	**17.30**
Ps	2	0.01	0.01	172.03	0.00	0.17
Em	2	5.27	2.64	81,364.46	0.00	**82.24**
Error	18	0.0001	0.00003			0.01
Total	26	6.41				100

Note Degrees of freedom (Df), sum of squares (SS), variance (V), F-value (F), P-value (P), percentage contribution (P (%))

of layers to be deposited in the stacking direction. For Sood et al. [25], a greater number of layers results in the formation of a high temperature gradient towards the base of the printed parts. This thermal process increases the efficiency of the diffusion mechanism between neighboring deposited filaments, improving the adhesion between them and, consequently, the mechanical strength of the parts.

Furthermore, the decrease in layer thickness compresses the extruded thermoplastic more intensively over the already deposited layers. The material passes from a cylindrical shape, given at its exit from the print nozzle to an oblong shape. This format increases the contact area, as well as the wettability, between the deposited structures, thus favoring a better adhesion process. Additionally, a tighter material leads to greater compaction between layers and reduced voids [26, 27].

The lower layer thickness might also have contributed to a more favorable thermal environment for the PLA crystallization process, promoting slow cooling and stable energy. Higher degree of crystallinity increases the mechanical properties of the polymer.

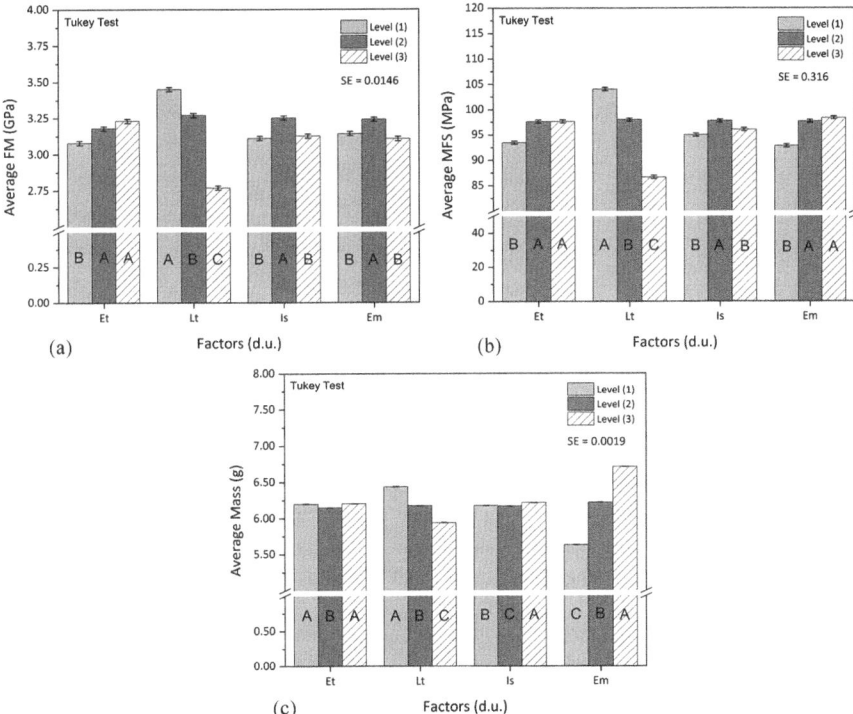

Fig. 5 Average by factors and levels for FM (**a**), MFS (**b**) and mass (**c**). PM filament

About the extrusion temperature, the higher its value, the greater the mechanical strength, Fig. 5a, b. However, the best values at 225 and 235 °C are high for PLA, which indicates that the polymer processing temperature could have been altered by the additives used in its formulation, including the red pigment, as well as by the thermal history of the filament extrusion process. These high levels of thermal energy are required to ensure fluidity to the material (facilitating extrusion), as well as power and stability to trigger the processes of bonding and neck growth between and within the layers, and the formation of the PLA crystalline structure.

For print speed, what is observed Fig. 5a, b—is that the best results for FM and MFS were for the central point, that is, the 60 mm/s setting. Possibly, this value is a balance between a good control of the volumetric flow with the thermal use and maintenance of the shape of the deposited filaments. A low speed, even more if combined with higher temperature levels, can induce thermal degradation to the material by exposure time. Peng and Wang [28] show that a low speed can lead to a process of burning material or destroying the layer by the hot nozzle. A high speed, on the other hand, can lead to the sharpening of the deposited filament.

The extrusion multiplier (Em) controls the amount of material, in filament length, that is being fed into the extrusion system [29]. As the value of "Em" increases, more

Table 5 Best levels by responses and verification model development, PM filament

Parameter (factor)	Response			Verification model (VM)
	FM	MFS	Mass	
Et	**235** or 225	**235** or 225	215 or **235**	**235**
Lt	0.1	0.1	0.1	0.1
Ps	60	60	80	60
Em	**1.0**	1.1 or 1.0	1.1	**1.0**

material is being supplied to the model to be printed, which is needed in this study with PM filament to improve the FM and MSF averages—Fig. 5a, b. This is probably a compensation for the material's fluidity deficiencies.

The mass of the parts, as expected (Table 4), was strongly affected by the extrusion multiplier (P (%) = 82.24) and layer thickness (P (%) = 17.30) factors. Both are related to the determination of the volume of material to be inserted in the model. A larger "Em", or a smaller layer thickness, tends to reduce the internal void density of objects in FFF, thus increasing their masses—Fig. 5c.

Theoretically, the mass[3] for a massive flexural sample in the PM PLA would be approximately 6.43 g. With a layer thickness of 0.1 mm, an average mass of 6.44 g was reached, that is, very close to the theoretical one. This behavior is related to the higher construction resolution obtained with the use of small thicknesses. The maximum level of "Em", 1.1, generated an average mass of 6.71 g, that is, an increase of 4% in relation to the stipulated value.

It is noteworthy, however, that the excess of material was not harmful to the quality of the surfaces and dimensional of the parts. The excess was well distributed, not disorganizing the meso-structure of the models, which can be seen by the good results generated in the maximum flexural strength tests.

Finally, Table 5 organizes the best levels of each parameter in each evaluated response and the selection process for the verification model.

The choice of layer thickness at 0.1 mm and print speed at 60 mm/s was simple, as a large number of correspondences (in green, Table 5) were found between the values as being the best among the responses obtained. In the case of temperature, statistical equality between two levels was always identified as the best result. The selection of the value of 235 °C for VM was based on the possible difficulty of flowing the filament and therefore the higher value would facilitate the reduction of the polymer viscosity and the 3D printing process. In addition, it would guarantee greater thermal energy to increase the mechanical resistance of the parts. The extrusion multiplier was set to 1 in the verification model as it met the criterion of improvement of a mechanical property, FM, and could balance mechanical strength and absence of excess material.

---

[3] Calculated from the ratio between the theoretical volume of the part and the measured density of the material.

3.2.2 KE Filament

Table 6 shows the ANOVA results for modulus, maximum flexural strength and mass of samples produced with the KE filament.

Regarding mechanical properties, all parameters, except for the FM print speed, were significant for changes in response behavior (Table 6). Despite the layer thickness having been a factor with considerable contribution to the variations of FM and MFS in the components manufactured with the KE filament, in this material, unlike in PM's, there is a division of protagonist with other parameters, more specifically the temperature and the extrusion multiplier.

The KE filament clearly requires a higher temperature level (235 °C), as shown in Tukey's analysis of Fig. 6a, b, to improve its mechanical strength. This characteristic indicates that it possibly has a low fluidity and, therefore, an increase in "Et" reduces the viscosity and allows the material to flow easily and homogeneously depending on the pressures exerted by the extrusion system.

Another evidence that configures the fluidity problem is the need for higher levels of "Em" (between 1 and 1.1.) to increase the values of the mechanical properties

Table 6 ANOVA for FM, MFS and mass responses, KE filament

Factor	ANOVA ($\alpha = 95\%$) for FM					
	Df	SS	V	F	P	P (%)
Et	2	0.23	0.11	20.38	0.00	**24.52**
Lt	2	0.38	0.19	34.30	0.00	**42.14**
Ps	(2)	(0.03)	*Pooled*			
Em	2	0.16	0.08	14.35	0.00	**16.89**
Error	18	0.08	0.005			16.45
Total	26	0.88				100
Factor	ANOVA ($\alpha = 95\%$) for MFS					
Et	2	143.78	71.890	72.13	0.00	**15.52**
Lt	2	349.80	174.900	175.47	0.00	**37.75**
Ps	2	22.24	11.12	11.16	0.00	2.40
Em	2	392.84	196.42	197.06	0.00	**42.40**
Error	18	17.94	0.997			1.94
Total	26	926.60				100
Factor	ANOVA ($\alpha = 95\%$) for Mass					
Et	2	0.02	0.01	61.09	0.00	0.21
Lt	2	1.22	0.61	4737.24	0.00	**16.46**
Ps	2	0.002	0.001	8.29	0.00	0.03
Em	2	6.17	3.08	23,970.78	0.00	**83.27**
Error	18	0.002	0.0001			0.03
Total	26	7.40				100

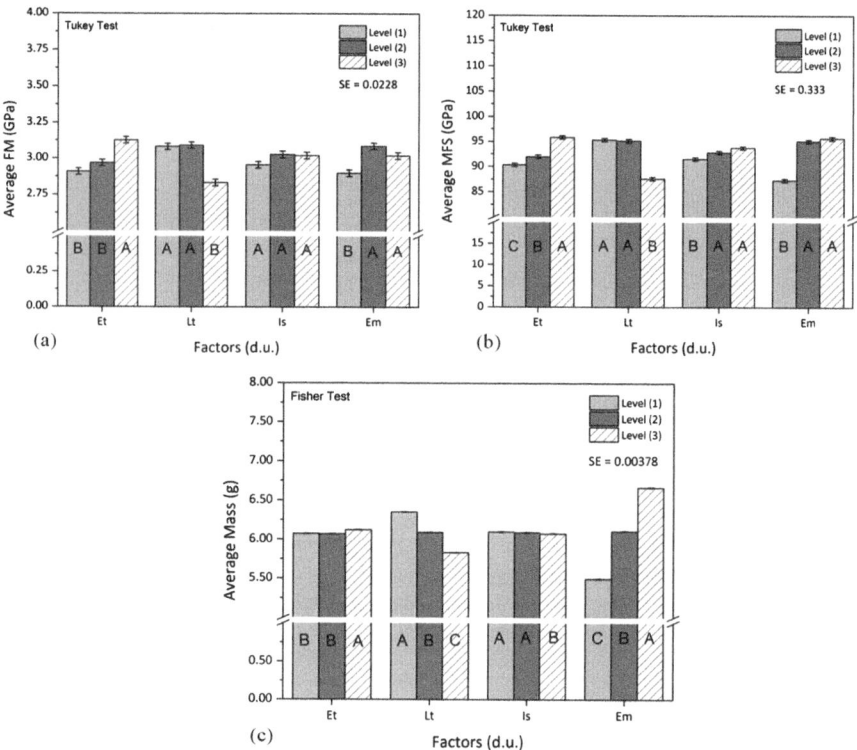

Fig. 6 Average by factors and levels for FM (**a**), MFS (**b**) and mass (**c**). KE filament

in flexural tests—Fig. 6a, b. When the polymer has low flowability, it resists extrusion, leading to filament slipping effects in the extruder head drive system, or even nozzle clogging. Slippage and clogging lead to material flow inconsistencies during deposition, impacting the dimensional and shape quality of the extrudate and, consequently, the adhesion between neighboring filaments deposited between and within the layers. To compensate for material losses, the length of filament supplied to the extruder is increased. This balances the amount of material needed to strengthen the union between the printed structures and reduce the void density of parts.

In summary, the parametric analysis of FM and MSF shows that the challenges of 3D printing with KE red PLA filament are in its processing, that is, in the phenomena that occur in the extrusion head.

The mass results, as in the PM, were strongly influenced by the extrusion multiplier (P (%) = 83.27) and by the layer thickness (P (%) = 16.46). The higher the "Em" and the lower "Lt", the greater the mass of the parts, Fig. 6c. As in the previous topic, the theoretical mass was calculated for a homogeneous component in KE, in this case 6.42 g. This value is slightly higher than that achieved with the mass average for layer thickness at 0.1 mm (6.35 g) and 4% lower than the average for "Em" at 1.1 (6.67 g).

Table 7 Best levels by responses and verification model development, KE filament

Parameter (factor)	Response			Verification model (VM)
	FM	MFS	Mass	
Et	235	235	235	235
Lt	**0.1** or 0.2	**0.1** or 0.2	**0.1**	**0.1**
Ps	–	80 or **60**	**60** or 40	**60**
Em	**1.1** or 1.0	**1.1** or 1.0	**1.1**	**1.1**

The synthesis of the best levels by responses as well as the verification model for the KE PLA, is presented in Table 7.

Developing the VM (Table 7) for the KE filament was not a simple task, since the only unanimous response (in green) of the parameter levels between FM, MFS, and mass was for the extrusion temperature at 235 °C. In the case of layer thickness, the value of 0.1 mm was selected considering that in FFF, the smaller the layer thickness, the greater the resolution of the process and the quality of the technological properties of the models. For the print speed, the level of 60 mm/s was chosen, as this appears as an option in MSF and in mass. Furthermore, this speed was chosen to compensate for the effects of the high extrusion temperature, i.e., avoiding thermal degradation of the material (40 mm/s) or stretching of the deposited filament (80 mm/s). Finally, the "Em" in 1.1 was adopted to compensate for any fluidity problems during extrusion.

3.2.3   F3D Filament

The F3D PLA filament was the great challenge of this research. Due to its low fluidity and sensitivity to temperature fluctuations, it offered many difficulties for processing in a conventional FFF 3D printing system. The problem was characterized by the sudden interruption of material deposition after the construction of a set of layers of the part. This failure resulted from clogging of the calibrated nozzle, caused, in turn, by the swelling of the filament at the nozzle/heating block interface (Fig. 7). Once the exit of the extrusion system was blocked, the filament could not move and a process of destruction (or machining) of the material began by the pulleys responsible for pulling it and pushing it towards the heated zone.

The material flow difficulty could be related to the absence, or low amount, of internal lubrication additives or to influences of the filament extrusion process. Often, to ensure diameter stability and decrease the size of production lines, the filaments are cooled (usually in water) quickly as they leave the extruder. This process induces the formation of cold crystallization effects in PLA, generated by the abrupt interruption of the polymer's natural crystallization process (slow cooling and with thermal stability). When reheated, the material restarts its crystallization, at low temperature levels, generating resistance to its deformation and, in this case, its extrusion and deposition.

Fig. 7 F3D filament
swelling

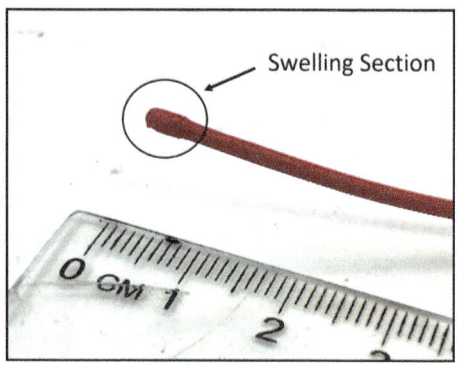

Among the combinations of the Taguchi experiment (Table 1), the one that presented the greatest difficulties for printing the F3D on the Prusa printer with a conventional extruder was Condition 1. In this condition, we have the lowest levels of extrusion temperatures (215 °C) and layer thickness (0.1 mm) tested. The lower the temperature, the higher the viscosity of the material and the lower its fluidity. Considering also thermal instabilities of the system, this temperature can vary below this level, further reducing the viscous transformation capacity of the polymer. A small layer thickness brings the nozzle very close to the layer already deposited, generating a back pressure effect that can impede the flow of material out of the extruder.

The Mosquito™ hotend, in turn, has greater capacity and thermal stability, which made it possible to overcome all the adversities imposed by the material or process parameters. However, the need for a special system already demonstrates the significant influence of the supplier on the quality of the material and its printability. In other words, not all PLA is easy to print, and the choice of supplier will be paramount for the FFF system user.

As for the study of process parameters, Table 8 presents the ANOVA results for FM, MFS, and mass.

ANOVA (Table 8) shows that all parameters influenced the mechanical responses, highlighting the importance of controlling the variables responsible for volumetric flow and amount of material, viscous transformation, and heat transfer mechanisms. The values of "Lt" (0.1 mm), "Ps" (60 mm/s) and "Em" (1.1) with better results in FM and MFS, Fig. 8a, b, agree with what was observed in the PM and KE filaments and therefore the justifications are compatible.

The extrusion temperature, however, presented a different scenario. The 225 °C level was the most efficient for improving the modulus and maximum flexural strength (Fig. 8a, b, respectively). The observed behavior could reflect the influence of the hardware element, the Mosquito™ hotend. A value of 235 °C, stable and constant, might have been high and capable of initiating a thermal degradation process in PLA. A value of 215 °C, once again, stable, and constant, might have evidenced the low fluidity of this material and the sensitivity to thermal fluctuations.

Table 8 ANOVA for FM, MFS, and mass responses, F3D filament

Factor	ANOVA (α = 95%) for FM					
	Df	SS	V	F	P	P (%)
Et	2	0.32	0.16	140.05	0.00	**18.62**
Lt	2	0.92	0.46	375.79	0.00	**53.80**
Ps	2	0.36	0.18	147.29	0.00	**21.09**
Em	2	0.09	0.04	36.32	0.00	5.20
Error	18	0.02	0.001			1.29
Total	26	1.71				100
Factor	ANOVA (α = 95%) for MFS					
Et	2	107.90	53.95	162.63	0.00	8.18
Lt	2	738.57	369.28	1113.22	0.00	**55.97**
Ps	2	137.56	68.78	207.34	0.00	**10.42**
Em	2	329.59	164.79	496.78	0.00	**24.98**
Error	18	5.97	0.332			0.45
Total	26	1319.59				100
Factor	ANOVA (α = 95%) for Mass					
Et	2	0.01	0.003	23.44	0.00	0.08
Lt	2	1.33	0.665	5396.10	0.00	**17.83**
Ps	2	0.01	0.004	28.44	0.00	0.09
Em	2	6.11	3.05	24,806.37	0.00	**81.97**
Error	18	0.02	0.0001			0.03
Total	26	7.45				100

A 10 °C change from 215 to 225 °C on a high-performance nozzle was enough to improve the viscous and mechanical response of the PLA F3D. In this sense, the thermal fluctuations of the conventional system are enough to cause clogging effects in the extruder head.

The mass follows the same pattern observed in the parts made with PM and KE filaments, with a strong influence of the extrusion multiplier and the layer thickness (Table 8). The highest average mass in "Em" of 1.1 (6.65 g), Fig. 8c, was approximately 3% greater than the theoretical mass of 6.48 g, calculated for the solid object in PLA F3D. The layer thickness of 0.1 mm provided an average mass of 6.36 g, i.e., 2% less than the theoretical one.

Table 9 provides a summary of the best parameter levels by response and global configuration (VM) formulation. The development of the verification model for the F3D filament showed the highest number of matches (in green, Table 9) between parameter levels and responses, which might be a consequence of the use of a more stable hotend.

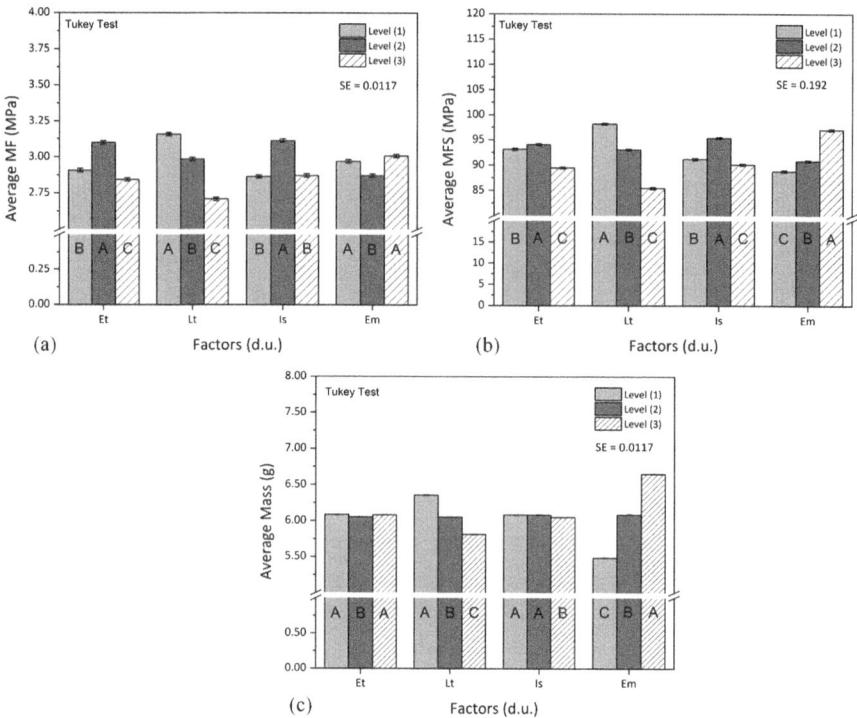

Fig. 8 Average by factors and levels for FM (**a**), MFS (**b**) and mass (**c**). F3D filament

Table 9 Best levels by responses and verification model building, F3D filament

Parameter (factors)	Response			Verification model (VM)
	FM	MFS	Mass	
Et	225	225	215 or 235	225
Lt	0.1	0.1	0.1	0.1
Ps	60	60	**60** or 40	60
Em	**1.1**	**1.1** or 0.9	**1.1**	**1.1**

3.2.4 Validation of Verification Models (VM)

In the graphs of Fig. 9, the averages of FM (a), MFS (b), and mass (b) are presented for the parts manufactured in each material with its verification model, as well as in each filament manufactured with the Castro [10] printing configuration.

The verification models were beneficial for the increase of the flexural modulus in the parts obtained with PM (+5%) and F3D (+4%) filaments in relation to those manufactured with these materials using the literature configuration—Fig. 9a. In the case of the KE, Figure (a), the VM did not change the rigidity of the parts in view

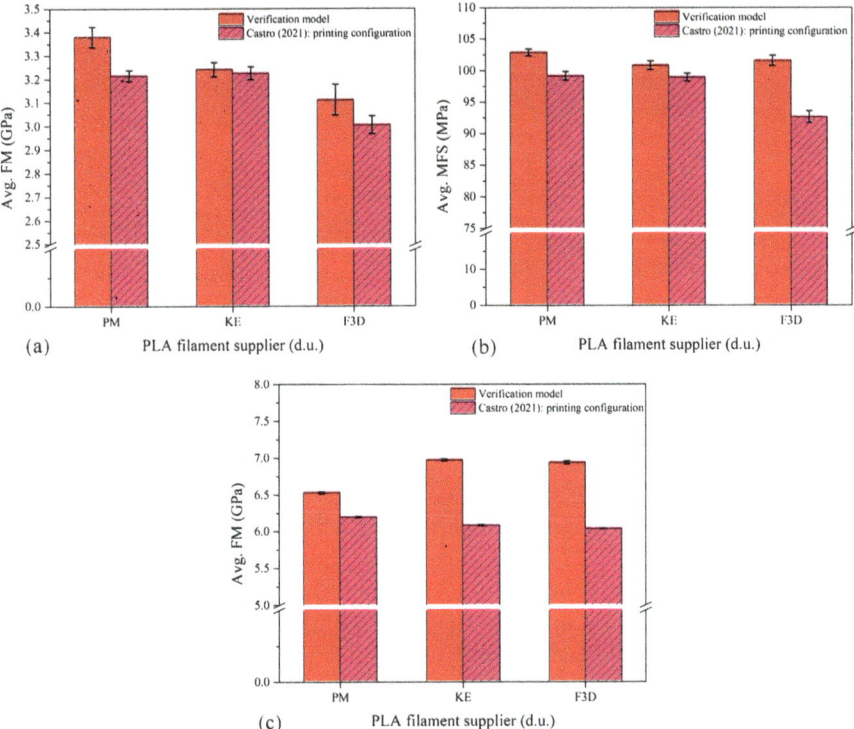

Fig. 9 Average FM (**a**), MFS (**b**), and mass (**c**) values for each filament manufactured with the verification model and with Castro's configuration [10]

of the results of applying the Castro [10] configuration to this filament. This result indicates the need for a process of refinement of the global configuration, mainly to improve the parameters that affect the temperature of the parts during the heating and cooling cycles of the FFF process and, consequently, the degree of crystallinity of the KE PLA.

In all cases, Fig. 9b, the maximum flexural stress increases from the literature fixed configuration for the filament-specific verification models. In PM the growth was 4% and in F3D it was 10%. The KE filament showed a 2% increase in MFS in the comparison performed, strengthening the need for a study to refine the parameters.

The mass of the parts increased considerably with the verification models. This result was expected, since the number of layers and the length of material fed in the parts built with VMs were significantly increased for the samples generated with the Castro [10] configuration. These differences were approximately +5% for PM and +15%, equally, for KE and F3D.

The models manufactured with the fixed setup had masses lower than their theoretical values for massive objects, being −4% for PM, −5% for KE, and −7% for F3D. With the verification model, the masses were higher than their theoretical value, being +2%, +9% and +7%, respectively for PM, KE, and F3D. It is therefore necessary to find a balance between material consumption and improvement of mechanical properties, which can be achieved with a parametric optimization study.

4   Conclusion

For many years, FFF 3D Printing was presented as a "plug-and-play" technology, that is, easy for anyone to have their machine at home and print quality parts consistently. However, this paper shows that a simple change of material supplier, or even the option for a specific color, can cause difficulties for users in their 3D printing system interaction. FFF is a technical manufacturing system and therefore its operation, even by enthusiasts, must be supported by adequate knowledge that must come from those who provide technologies and materials for the process.

Research with different manufacturers of red filament made it possible to identify significant differences in material properties. Such results indicate a promising way to understand the relationship of printing parameters by the FFF technique. These observations allow an entry level user to better understand the different print results for the same material from different suppliers. Although the study was carried out with PLA, the methodology developed, and its interpretations, can be applied with other materials.

In the opposite way, it is possible that such analysis allows a better perception of the manufacturers for the care with the formulation of the materials and the relationship of the pigment and additives in the blend. It is a recommendation of this study that manufacturers develop a manual of good printing practices with the procedures and parameters for their specific materials. The user, in turn, must develop routine of reading the recommendations and carrying out a sequence of tests to develop the best configuration in their printing system.

The results of the parametric analysis show that the red pigment changes the thermal behavior of PLA, either in terms of processing or mechanical responses of the printed parts. This confirmation contributes to answer the hypothesis mentioned by Castro [10], which said that red materials would need higher temperatures for viscous transformation. In addition, different filaments suppliers can apply different additives that change the fluidity of the material, which requires a specific parameterization study and, in extreme cases, the adaptation of the 3D printing systems.

The filament characterization technique corresponded to the data indicated by the manufacturers, especially in density. The FTIR also indicated that the filaments, in their molecular structure, correspond to the PLA material. The main difference between suppliers was the response of the printed parts to the selected parameters and to the flexural tests, indicating that the filaments need different processing. Verification models, i.e., global print setup, improved the quality of printed parts for different suppliers. However, a new round of studies is needed to optimize the parameters to balance good fluidity, good mechanical properties, and mass reduction.

For future work, it will be necessary to further investigate the thermal, chemical, and rheological differences of filaments using Differential Scanning Calorimetry (DSC), Thermogravimetric Analysis (TGA) and Chromatography and Measurement of the Fluid Index (MFI).

Acknowledgements FEUP and INEGI for providing the equipment and resources to carry out this research. To Professor Guilherme Barra, UFSC/Brazil, for providing the DMA equipment and the weighing scale for measuring material density.

References

1. Tang, C., Seeger, S.: Systematic ranking of filaments regarding their particulate emissions during fused filament fabrication 3D printing by means of a proposed standard test method. Indoor Air **32**, e13010 (2022). https://doi.org/10.1111/ina.13010
2. Antoniac, I., Popescu, D., Zapciu, A., Antoniac, A., Miculescu, F., Moldovan, H.: Magnesium filled polylactic acid (PLA) material for filament based 3D printing. Materials (Basel) **12**, 719 (2019). https://doi.org/10.3390/ma12050719
3. Song, H., Martínez, J., Bedell, P., Vennin, N., Lefebvre, S.: Colored fused filament fabrication. ACM Trans. Graph. **38**, 1–11 (2019). https://doi.org/10.1145/3183793
4. Wang, Z., Shen, H., Wu, S., Fu, J.: Colourful fused filament fabrication method based on transitioning waste infilling technology with a colour surface model. Rapid Prototyp. J. **27**, 145–154 (2021). https://doi.org/10.1108/RPJ-04-2020-0072
5. Wittbrodt, B., Pearce, J.M.: The effects of PLA color on material properties of 3-D printed components. Addit. Manuf. **8**, 110–116 (2015). https://doi.org/10.1016/j.addma.2015.09.006
6. Davis, A.Y., Zhang, Q., Wong, J.P.S., Weber, R.J., Black, M.S.: Characterization of volatile organic compound emissions from consumer level material extrusion 3D printers. Build. Environ. **160**, 106209 (2019). https://doi.org/10.1016/j.buildenv.2019.106209
7. Hanon, M.M., Zsidai, L., Ma, Q.: Accuracy investigation of 3D printed PLA with various process parameters and different colors. Mater. Today Proc. **42**, 3089–3096 (2021). https://doi.org/10.1016/j.matpr.2020.12.1246
8. Matos, B.D.M., Rocha, V., da Silva, E.J., Moro, F.H., Bottene, A.C., Ribeiro, C.A., dos Santos Dias, D., Antonio, S.G., do Amaral, A.C., Cruz, S.A., de Oliveira Barud, H.G., Silva Barud, H. da: Evaluation of commercially available polylactic acid (PLA) filaments for 3D printing applications. J. Therm. Anal. Calorim. **137**, 555–562 (2019). https://doi.org/10.1007/s10973-018-7967-3
9. Bermudez, D., Quiñonez, P.A., Vasquez, E.J., Carrete, I.A., Word, T.J., Roberson, D.A.: A comparison of the physical properties of two commercial 3D printing PLA grades. Virtual Phys. Prototyp. **16**, 178–195 (2021). https://doi.org/10.1080/17452759.2021.1910047
10. Castro, F.F.A. de: Análise da influência da pigmentação na qualidade de peças impressas por FFF em PLA e PETG. https://hdl.handle.net/10216/135101 (2021)
11. ASTM International: Standard test methods for flexural properties of unreinforced and reinforced plastics and electrical insulating materials. United States of America (2010)
12. Santana, L., Lino Alves, J., da Costa Sabino Netto, A., Merlini, C.: Estudo comparativo entre PETG e PLA para Impressão 3D através de caracterização térmica, química e mecânica. Rev. Matéria. **23**, e12267 (2018). https://doi.org/10.1590/s1517-707620180004.0601
13. Santana, L., Lino Alves, J., da Costa Sabino Netto, A.: Dimensional analysis of PLA and PETG parts built by open source extrusion-based 3d printing. In: 10° Congresso Brasileiro de Engenharia de Fabricação. ABCM (2019)

14. Santana, L., Lino Alves, J., da Costa Sabino Netto, A.: A study of parametric calibration for low cost 3D printing: seeking improvement in dimensional quality. Mater. Des. **135**, 159–172 (2017). https://doi.org/10.1016/j.matdes.2017.09.020
15. Rocca-Smith, J.R., Lagorce-Tachon, A., Iaconelli, C., Bellat, J.P., Marcuzzo, E., Sensidoni, A., Piasente, F., Debeaufort, F., Karbowiak, T.: How high pressure CO2 impacts PLA film properties. Express Polym. Lett. **11**, 320–333 (2017). https://doi.org/10.3144/expresspolym lett.2017.31
16. Pop, M.A., Croitoru, C., Bedő, T., Geamǎn, V., Radomir, I., Cosniţǎ, M., Zaharia, S.M., Chicos, L.A., Milosan, I.: Structural changes during 3D printing of bioderived and synthetic thermoplastic materials. J. Appl. Polym. Sci. **136**, 47382 (2019). https://doi.org/10.1002/app. 47382
17. Arora, J.K., Bhati, P.: Fabrication and characterization of 3D printed PLA scaffolds. In: Cebeci, F.C., Menceloglu, Y.Z., Unal, S., Ulcer, Y. (eds.) AIP Conference Proceedings 2205, pp. 020065-1–020065-5. AIP Publishing, Turkey (2020)
18. Haryńska, A., Janik, H., Sienkiewicz, M., Mikolaszek, B., Kucińska-Lipka, J.: PLA–Potato thermoplastic starch filament as a sustainable alternative to the conventional PLA filament: processing, characterization, and FFF 3D printing. ACS Sustain. Chem. Eng. **9**, 6923–6938 (2021). https://doi.org/10.1021/acssuschemeng.0c09413
19. Filament PM: Technical data sheet for product: PLA filament. https://shop.filament-pm.com/ pla-red-1-75-mm-1-kg/p70 (2022)
20. Kexcelled: Technical data sheet: Kexcelled PLA K5. https://kexcelled.nl/product/pla-k-5 (2020)
21. Tucab: Fil3D PLA 3D850: Ficha técnica. https://www.tucab.pt/pt/Filamentos-para-Impressao-3D/Filamento-3D-PLA-3D850 (2021)
22. Henton, D., Gruber, P., Lunt, J., Randall, J.: Polylactic acid technology. In: Mohanty, A.K., Misra, M., Drzal, L.T. (eds.) Natural Fibers, Biopolymers, and Biocomposites, pp. 527–578. CRC Press, Boca Raton (2005)
23. Patti, A., Acierno, S., Cicala, G., Zarrelli, M., Acierno, D.: Assessment of recycled PLA-based filament for 3D printing. In: Cicala, G., Díez-Pascual, A.M., Yusa, S. (eds.) The 2nd International Online Conference on Polymer Science—Polymers and Nanotechnology for Industry 4.0, p. 16. MDPI, Basel Switzerland (2021)
24. Vigil Fuentes, M.A., Thakur, S., Wu, F., Misra, M., Gregori, S., Mohanty, A.K.: Study on the 3D printability of poly(3-hydroxybutyrate-co-3-hydroxyvalerate)/poly(lactic acid) blends with chain extender using fused filament fabrication. Sci. Rep. **10**, 11804 (2020). https://doi.org/10. 1038/s41598-020-68331-5
25. Sood, A.K., Ohdar, R.K., Mahapatra, S.S.: Parametric appraisal of mechanical property of fused deposition modelling processed parts. Mater. Des. **31**, 287–295 (2010). https://doi.org/ 10.1016/j.matdes.2009.06.016
26. Syrlybayev, D., Zharylkassyn, B., Seisekulova, A., Akhmetov, M., Perveen, A., Talamona, D.: Optimisation of strength properties of FDM printed parts—a critical review. Polymers (Basel) **13**, 1587 (2021). https://doi.org/10.3390/polym13101587
27. Lokesh, N., Praveena, B.A., Sudheer Reddy, J., Vasu, V.K., Vijaykumar, S.: Evaluation on effect of printing process parameter through Taguchi approach on mechanical properties of 3D printed PLA specimens using FDM at constant printing temperature. Mater. Today Proc. **52**, 1288–1293 (2022). https://doi.org/10.1016/j.matpr.2021.11.054
28. Peng, A.H., Wang, Z.M.: Researches into influence of process parameters on FDM parts precision. Appl. Mech. Mater. **34–35**, 338–343 (2010). https://doi.org/10.4028/www.scientific.net/ AMM.34-35.338
29. Pivar, M., Gregor-Svetec, D., Muck, D.: Effect of printing process parameters on the shape transformation capability of 3D printed structures. Polymers (Basel) **14**, 117 (2021). https:// doi.org/10.3390/polym14010117

Machining

Calibration of Grinding Wheel with On-Line Acquisition of Grinding Force

Lai Hu, Jun Zha, and Yaolong Chen

Abstract Aiming to optimize the grinding parameters accurately and improve the grinding accuracy of parts, the wheel of on-line grinding force measuring was innovatively designed. Under the same working conditions, different weights and calibrations were used to study the precision calibration. Meanwhile, Wheatstone Bridge test method was used to analyze the calibration accuracy in vertical direction and $90°$ rotation direction. The results show that the output voltage increases with the weight of the weight, whether it was vertical test or rotating $90°$ test. Under the same weight, the output voltage of vertical test was larger than that of rotating $90°$, and the output voltage of vertical test was about twice that of rotating $90°$. Based on this research, the traditional indirect measurement of grinding force was directly changed into direct measurement, which improves the measurement accuracy of grinding force.

Keywords Grinding parameters · Grinding force · Calibration · Measurement

1 Introduction

Precision grinding is one of the most important processes in high-end parts. The grinding force determines the microscopic changes of residual stress and retained austenite content on the surface of parts. Therefore, we need to accurately collect the

L. Hu · J. Zha (✉) · Y Chen
School of Mechanical Engineering, Xi'an Jiaotong University, 28 Xianning Road, Xi'an, Shaanxi 710049, P.R. China
e-mail: jun_zha@xjtu.edu.cn

L. Hu
e-mail: hulai0405@stu.xjtu.edu.cn

Y. Chen
e-mail: chenzwei@mail.xjtu.edu.cn

State Key Laboratory for Manufacturing Systems Engineering, Xi'an Jiaotong University, Xi'an, Shaanxi 710054, P.R. China

© The Author(s), under exclusive license to Springer Nature Switzerland AG 2023
L. F. M. da Silva et al. (eds.), *1st International Conference on Engineering Manufacture 2022*, Proceedings in Engineering Mechanics,
https://doi.org/10.1007/978-3-031-13234-6_4

contact grinding force between grinding wheel and workpiece. The grinding parameters are optimized and controlled by grinding force. At present, some authors have studied the grinding force. Dai [1] studied the role of grinding force in the formation mechanism of white layer deeply and systematically by means of experimental analysis, theoretical derivation and numerical simulation. The plastic deformation caused by grinding force and the influence of contact stress on the formation of white layer were analyzed. The optimum values of grinding parameters were obtained when the thickness of white layer is minimum. Su et al. [2] calculated the heat distribution ratio and convection heat dissipation coefficient during grinding. Zhang et al. [3] studied grinding process parameters (including cutting depth, feed rate and spindle speed) through temperature and force sensors. The results shown that cutting depth has the greatest influence on grinding temperature and normal force. Ruzzi et al. [4] tested five different cutting conditions. The grinding performance was evaluated by surface smoothness and cutting zone temperature. Ding et al. [5] established the theoretical model of grinding force of grinding wheel. The rail grinding process was simulated by DEFORM-3D software. The influence of grinding parameters on grinding force was discussed. The simulation and experimental results show that the grinding force increases with the increase of grinding pressure and wheel size, and decreases with the increase of wheel speed. The error between simulation results and experimental results was reduced from 10.22 to 4.42%. There were many researches on grinding force and grinding temperature [6–9].

This study mainly improved the traditional grinding wheel structure. The acquisition scheme of grinding force was made by embedding strain gauge. The calibration of grinding force grinding wheel was studied by using Wheatstone bridge theory.

2   Design and Calibration Theory Analysis of Grinding Wheel for On-Line Acquisition of Grinding Force

2.1   Design of Grinding Wheel for On-Line Grinding Force Collection

The traditional grinding wheel matrix was improved, and the structure was shown in Fig. 1.

In Fig. 1, where 1-CBN block abrasive sheet, 2-lock the bolt, 3-grinding wheel matrix, 4-grinding wheel cover plate, 5-lock the bolt, 6-basal body of the abrasive wheel and the circumferential viscose layer of the CBN block abrasive sheet, 7-glue layer of grinding wheel substrate and CBN block abrasive sheet section, 8-wire hole of strain gauge of tangential force (X) collect, 9-installation groove of strain gauge of tangential force (X) collect, 10-strain gauge of tangential force (X) collect, 11-strain gauge of axial force (Y) collect, 12-wire hole of strain gauge of axial force (Y) collect, 13-strain gauge of Z direction force collect, 14-installation groove of strain gauge of Z direction force collect, 15-installation groove of strain gauge of axial

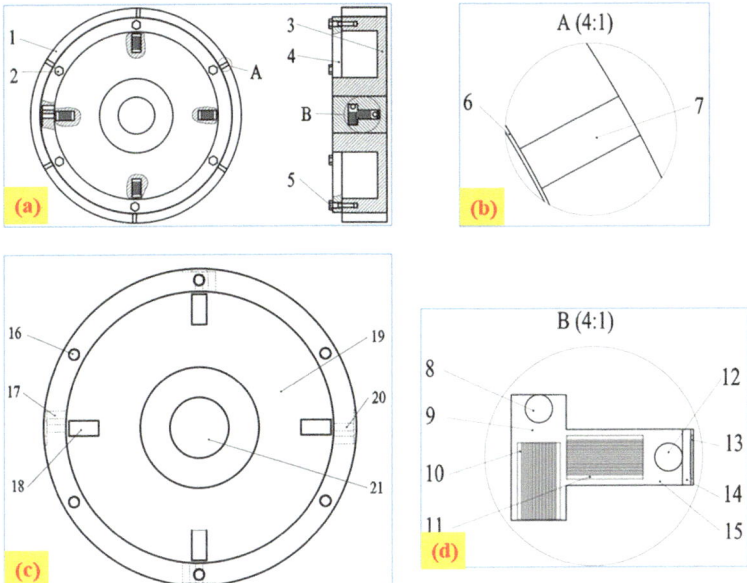

Fig. 1 Grinding wheel structure with wireless acquisition of grinding force: **a** the overall structure of the grinding wheel, **b** the enlarged drawing of part A, **c** the internal structure of the grinding wheel, and **d** the strain gauge sticking structure

force (Y) collect, 16-cover plate bolt installation hole, 17-wire hole of strain gauge of tangential force (X) collect, 18-installation groove of strain gauge of Z direction force collect, 19-installation groove of wireless module of force collect, 20-wire hole of strain gauge of axial force (Y) collect, 21-installation hole of grinding wheel shank.

The whole structure of grinding wheel base includes three blocks: grinding wheel base 3, CNB block abrasive 1 and grinding wheel cover 4. The scheme structure of the force acquisition and transmission module comprises force strain gauges (10, 11 and 13), a signal acquisition module, a signal amplification module and a signal wireless transmission module.

Specifically, the processed grinding wheel substrate 3 was cleaned. In particular, a thin layer of fine sand was sprayed at the force collecting strain gauge mounting grooves (9, 15 and 14) in the tangential (X), axial (Y) and Z directions of the grinding wheel base for strain gauge bonding. Thereafter, the tangential (X), axial (Y), and Z-direction force collecting strain gauges were properly attached to the mounting grooves 9, 15, and 14, respectively, and cooled for 3 min. The signal wires of the tangential (X) force acquisition strain gauge 10 and the axial (Y) force acquisition strain gauge 11 were penetrated into the grinding wheel base force acquisition wireless module mounting groove 19 through the tangential (X) force acquisition strain gauge wire hole 17 and the axial (Y) force acquisition strain gauge wire hole 20, respectively.

The processed CBN block abrasive sheet 1 was bonded on the circumference of the grinding wheel step by step through an adhesive, and a adhesive layer 6 on the circumference of the grinding wheel substrate and the CBN block abrasive sheet was formed. There was a certain gap between each CBN block abrasive sheet 1, and a cross-sectional viscose layer was formed between the grinding wheel substrate and the CBN block abrasive sheet, which was heated at 50 °C and kept for 3 h.

The completed wireless transmission module was placed in a grinding wheel base force acquisition wireless module mounting groove 19 and connected with wires of a tangential (X) force acquisition strain gauge 10, an axial (Y) force acquisition strain gauge 11 and a Z-direction force acquisition strain gauge 13.

The grinding wheel cover plate 4 was connected and locked with the grinding wheel base body 3 through the six cover plate bolt mounting holes 16 through the locking bolts (2 and 5).

The whole structure of the grinding wheel was mounted with the grinding wheel through the grinding wheel shank mounting hole 21 and the grinding force signal was collected.

2.2 Theoretical Analysis of Grinding Force Calibration

Wheatstone Bridge is an instrument that can accurately measure the resistance. It uses the change of resistance to measure the change of physical quantity. The function of single chip microcomputer is to collect the voltage at both ends of variable resistance, and then calculate the corresponding change of physical quantity through the processed data. The resistors R1, R2, R3 and R4 constitute the four arms of the bridge, and G is a galvanometer, which is used to check whether there is current in the branch where it is located. The schematic diagram of the specific structure is shown in Fig. 2.

Because the resistor voltages of R_1, R_2 and R_3, R_4 are different, the bridge may be unbalanced. Calculate as Eq. (1):

Fig. 2 Wheatstone bridge

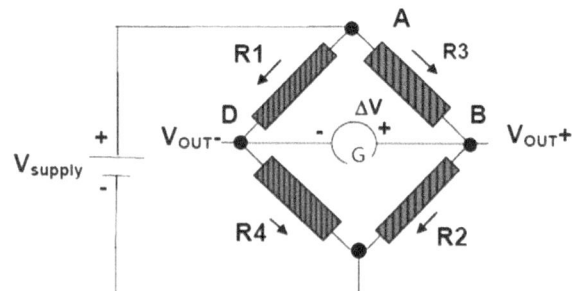

$$V_0 = V_s \left(\frac{R_1}{R_1 + R_2} - \frac{R_4}{R_3 + R_4} \right) \tag{1}$$

If the "bridge" is balanced and $\frac{R_1}{R_2} = \frac{R_3}{R_4}$, the bridge output voltage V_0 is zero.

Under the condition of preset strain, the resistance change of strain gauge ΔR. Therefore, Eq. (2)

$$V_0 = V_s \left(\frac{R_1 + \Delta R_1}{R_1 + \Delta R_1 + R_2 + \Delta R_2} - \frac{R_4 + \Delta R_4}{R_3 + \Delta R_3 + R_4 + \Delta R_4} \right) \tag{2}$$

For strain measurements, the resistances R_1 and R_2 must be equal in the Wheatstone Bridge. The same applies to R_3 and R_4. Therefore, this theory was used as the calibration method of grinding force.

3   Grinding Force Calibration Analysis

The calibration test bench was established by Fig. 2 and Wheatstone bridge theory, as shown in Fig. 3.

According to the grinding force calibration test bench in Fig. 3, vertical calibration and rotation 90° calibration were carried out respectively. At the same time, different

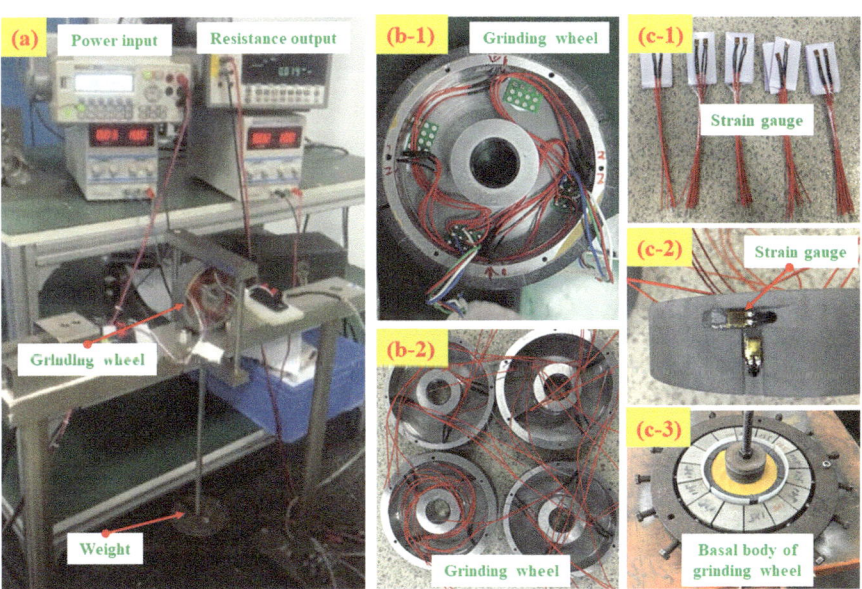

Fig. 3 Grinding force calibration experiment

weights (2, 4, 6, 8, 10, 12, 14, 20, 30, 40 and 50 kg) were compared and analyzed, as shown in Figs. 4 and 5.

According to the macroscopic analysis of Figs. 4 and 5, the output voltage gradually increased with the increase of weight, whether it was vertical test or rotating 90° test. Under the same weight, the output voltage of vertical test was larger than that of rotating 90°.

From the microscopic analysis of Figs. 4 and 5, when the weight of the weight was 2 kg, the output voltages of vertical test and rotation 90° test were 0.009 V and 0.004 V, respectively. When the weight of weight was 10 kg, the output voltages of vertical test and rotation 90° test were 0.039 V and 0.019 V, respectively. When the weight of weight was 50 kg, the output voltages of vertical test and rotation 90° test were 0.171 V and 0.084 V, respectively. From the analysis of these three sets of data, the vertical test output voltage was about twice that of the rotation 90° test voltage under the same test weight.

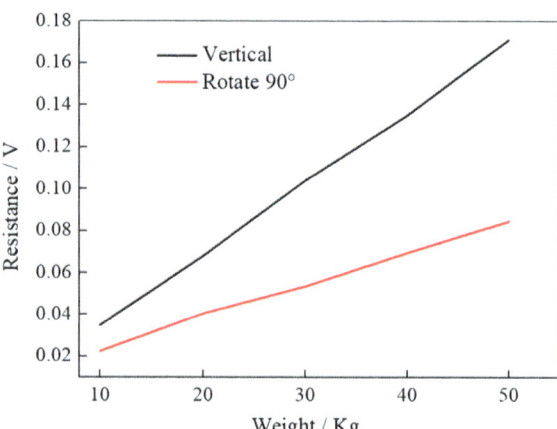

Fig. 4 Grinding force calibration with weights of 10, 20, 30, 40 and 50 kg

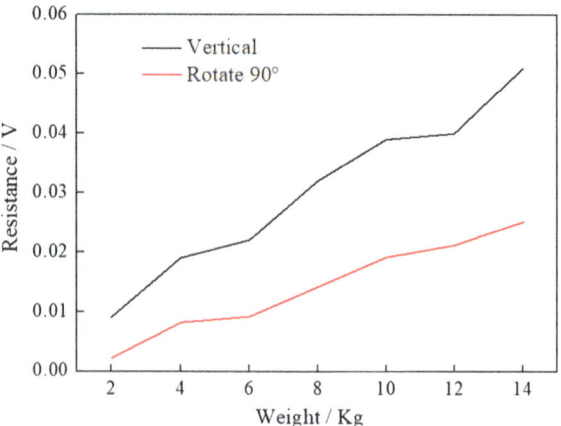

Fig. 5 Grinding force calibration with weights of 2, 4, 6, 8, 10, 12 and 14 kg

4   Conclusion

In this study, the structure of traditional grinding wheel was improved, and strain gauges were embedded in it. Through strain gauge, the grinding force was calibrated and analyzed by wireless acquisition.

(a) In the calibration, the Wheatstone Bridge theory was analyzed, and the Wheatstone Bridge test-bed was designed and developed.
(b) Calibration of grinding force of grinding wheel was carried out by Wheatstone theory and test bench. Under the same test weight, the output voltage of vertical test was larger than that of rotating 90°. The vertical test output voltage was about twice that of the rotation 90° test voltage. At the same time, it was also analyzed that the output voltage gradually increases with the weight of the weight, whether it was vertical test or rotating 90° test.
(c) In the following work, grinding test and calibration results will be analyzed to accurately obtain the grinding force when the grinding wheel contacts with the workpiece, and the adaptive optimization of grinding parameters will be studied.

CRediT authorship contribution statement

Lai Hu: Conceptualization, Methodology, Formal analysis, Writing—original draft. **Jun Zha**: Conceptualization—review and editing. **Yaolong Chen**: Supervision.

Declaration of competing interest

The authors declare that they have no known competing financial interests or personal relationships that could have appeared to influence the work reported in this paper.

Acknowledgements This research was funded by the National Key R&D Program of "Manufacturing Basic Technology and Key Components" (No. 2020YFB2009604 and No. 2018YFB2000502).

References

1. Dai, M.: Research on the Effect of Grinding Force on Mechanism of White Layer Formation in Grinding of Quench Bearing Steels. Hunan University
2. Su, J.X., Zhang, Y.Z., Deng, X.Z.: Analysis and experimental study of cycloid gear form grinding temperature field. Int. J. Adv. Manuf. Technol. **110**(3–4), 1–17 (2020)
3. Zhang, L., Zou, L., Wen, D., Wang, X., Piao, Z.: Investigation of the effect of process parameters on bone grinding performance based on on-line measurement of temperature and force sensors. Sensors **20**(11), 3325 (2020)
4. Ruzzi, R., Paiva, R., Machado, A.R., Silva, R.B.D.: Analysis of temperature and surface finish of Inconel 718 during grinding utilizing different grinding wheels. J. Brazil. Soc. Mech. Sci. Eng. **43**(5) (2021)

5. Ding, H.H., Han, Y.C., Zhou, K.: Grinding force modeling and experimental verification of rail grinding. ARCHIVE Proc. Inst. Mech. Eng. Part J: J. Eng. Tribol. 1994–1996 (vols. 208–210) **234**(8), 135065011990073 (2020)
6. Pdw, A., Hee, B., Cbp, B.: Investigation into the wear behaviour of Stellite 6 during rotation as an unlubricated bearing at 600 °C. Tribol. Int. (ScienceDirect) **44**(12), 1589–1597 (2011)
7. Deguang, L.I.: Normal grinding force measuring based on high speed magnetic bearing spindle. J. Mech. Eng. **48**(24), 1 (2012)
8. Gu, J., Zhang, Y., Liu, H.: Influences of wear on dynamic characteristics of angular contact ball bearings. Meccanica **54**(7), 945–965 (2019)
9. Jamshidi, H., Budak, E.: An analytical grinding force model based on individual grit interaction. J. Mater. Process. Technol. **283**, 116700 (2020)

Injection

Improving the Efficiency of the Bowden Cable Terminal Injection Process for the Automotive Industry

J. L. T. A. Pereira, R. D. S. G. Campilho⊙, F. J. G. da Silva⊙, and I. J. Sánchez-Arce⊙

Abstract Control cables allow the transfer of force between two separate locations by a flexible means, hence their importance in the automotive industry and many others; their terminals interact with those moving and moved mechanisms, so must be strong. The cable terminals are created by injection moulding which has the inherent requirement of leaving extra material (sprues) to allow the correct mould filling and to compensate for the dimensional changes during the cooling process. Sprues should be removed in a controlled way and disposed for recycling. However, features within the moulds may lead to larger sprues, porosity in the terminals, and a brittle interface between the terminal and the sprue. The latter causes early sprue separation, causing other issues in the production line. This work addresses the sprue size and its consequences from an industrial case study. In this case, the process was studied following a systematic approach to establishing the most effective solution. Subsequently, the mould and mould structure were improved, allowing for a controlled sprue size and a common mould structure for 95% of the products of the company, which reduced the tooling preparation time. Then, adjacent components were addressed to facilitate the product flow. The improvements were minimal invasive, keeping compatibility with the componentry of other production lines. The modification led to a reduction of 62.6% in the sprue mass, porosity was reduced by 10.2 and 55.9%, corresponding to two terminal models. In conclusion, the interventions done fulfilled the requirements and improved the operation of the line.

J. L. T. A. Pereira · R. D. S. G. Campilho (✉) · F. J. G. da Silva
Departamento de Engenharia Mecânica, Instituto Superior de Engenharia do Porto, Instituto Politécnico do Porto, R. Dr. António Bernardino de Almeida, 431, 4200-072 Porto, Portugal
e-mail: raulcampilho@gmail.com

J. L. T. A. Pereira
e-mail: joseluis_pereira@outlook.pt

F. J. G. da Silva
e-mail: fgs@isep.ipp.pt

R. D. S. G. Campilho · F. J. G. da Silva · I. J. Sánchez-Arce
INEGI – Pólo FEUP, Rua Dr. Roberto Frias, 400, 4200-465 Porto, Portugal
e-mail: isidrodjsa@gmail.com; iarce@inegi.up.pt

© The Author(s), under exclusive license to Springer Nature Switzerland AG 2023
L. F. M. da Silva et al. (eds.), *1st International Conference on Engineering Manufacture 2022*, Proceedings in Engineering Mechanics,
https://doi.org/10.1007/978-3-031-13234-6_5

Keywords Control cables · ZAMAK injection · Die casting · Finite element method · Process improvement

1 Introduction

In the global scenario of the industry, all organizations are dedicated to being more competitive with the increase in productivity achieved through superior quality manufacturing processes and at a lower cost [1]. Quality control has the function of guaranteeing and improving the quality of the product through inspection and process control [2]. Ishikawa [3] proposed seven basic tools to find solutions to problems and achieve process improvements. Continuous quality improvement is a management strategy that aims to maintain and improve quality through constant evaluation of causes that generate quality defects. After identifying the causes, solutions are created to avoid defects and make optimizations [4]. Tools like brainstorming, Ishikawa (or cause and effect) diagrams, and Strengths, Weaknesses, Opportunities, and Threats (SWOT) analyses are particularly useful for process improvement in industrial environments [5].

Brainstorming is a process used to generate ideas in a non-judgmental environment [4]. There are three fundamental principles for brainstorming [6]: (1) aim for quantity instead of the quality of ideas, (2) freedom of thought, and (3) encourage the new and innovative. Typically, this technique involves the following steps: problem presentation, idea creation, idea review, idea selection, and idea prioritization. For example, Rodrigues et al. [7] used this technique to solve the mismanagement of waste reports that existed in a company in the automotive industry by means of a computer application developed by them. The brainstorming involved the company's employees, leading to the application's requirements. The proposed solution led to an increase in the reliability of waste reports by 76% and reduce their associated cost and time by 75%. On the other hand, a cause-effect diagram is a tool used for identifying the possible causes of a problem [4]. The causes are organized by families, which in turn are divided into subcategories. The families are the method, labour, raw material, measurement and environment [3]. In this regard, Silva et al. [8] carried out a study to reduce finishing operations and obtain defect-free parts in an injection moulding process. Cause-effect diagrams were used to analyse all factors influencing the quality of the produced parts, guidelines were established which led to the intended results within the process. Likewise, the SWOT analysis is a strategic quality tool to analyse both internal and external factors involved in a process. Through external analysis, the threats and opportunities of the competitive environment in which the process operates are identified, while the internal analysis helps to assess the organization's strengths and weaknesses [9]. After identifying these factors, strategies are developed in order to intensify strengths, eliminate weaknesses, explore opportunities and combat threats [10]. For example, Karimi et al. [11] employed this technique to draw a business strategy for a ceramic sector industry. The objectives were met, and the SWOT technique proved to be effective for these analyses.

Metal casting is a widely used process for large production runs due to the low cost and flexibility in the components' geometry and materials. The most common alloys used for metal casting are iron, aluminium, steel, and copper [12]. Despite the advantages of metal casting, issues related to final mechanical properties, porosity, surface finish, and dimensional accuracy are often discussed [13, 14]. Die casting, a variant of metal casting, is characterized by filling the mould cavity by injecting molten metal at high speed and pressure (7–350 MPa). The mould is open once the metal has solidified [14], resulting in parts with good mechanical properties, high dimensional accuracy, and excellent surface finishes [8, 15]. It is a process of extreme precision since, otherwise, working at high temperatures and pressures in short time intervals would cause numerous problems [16]. The steel moulds are made up of two halves, to allow the extraction of the parts, and enable serial production of parts. The process begins with the closing of the moulds, followed by the injection of molten metal. After solidification, the mould is opened, and the parts are extracted. Improvement of the die casting process has been researched [17, 18]. For example, Nunes et al. [19] studied the wear mechanisms of two moulds with severe problems. Mould coating was proposed, and its benefits were tested by analysing the wear resistance behaviour and respective wear mechanisms. After systematic analyses and testing, it was found that the $Ti_{40}Al_{60}N$ coating provided the lowest friction coefficient and improved wear resistance. Later, the injection parameters necessary to produce a ZAMAK part were optimized by Silva et al. [8]. The part had to be injected in one shot and should possess a good surface finish. The optimization was carried out using computational software (SolidCast™) in which the metal flow within the mould was analysed, leading to improving the mould design. In this regard, turbulence during the injection process may influence the part's porosity, as demonstrated by Sun et al. [20] through numerical simulations.

This work addresses the sprue size and its consequences from an industrial case study, which is described in Sect. 2. In this case, the process was studied; also, techniques such as the Strengths, Weaknesses, Opportunities, and Threats (SWOT) method together with the Finite Element Method (FEM) were employed to determine the best alternative solutions. Subsequently, the mould and mould structure were improved, allowing for a controlled sprue size and a common mould structure for 95% of the products of the company, which reduced the tooling preparation time. Then, adjacent components like the measuring kit, the strength testing device, and the structure of the robotic manipulator were addressed to facilitate the product flow. All the interventions were planned and as minimally invasive as possible whilst taking advantage of the components currently in operation, maintaining the compatibility with the other production lines in the company and reducing the interventions' costs.

2 Problem Description

The production line for Bowden cables analysed here is part of a company located in northern Portugal whilst the analysis and design were done under their request; the company name is kept anonymous by their request. One of the processes performed in the production line is the injection moulding of cable terminals suiting several applications; thus, moulds are specific for each model. The injection moulding process is briefly described below.

2.1 Process Characterization

The fabrication of a control cable starts with the injection of the first terminal to the inner cable, done in another production line. Then, the inner cable is transferred to line BX726, the objective of this study. Currently, line BX726 produces four different models of control cables. The preparation process is model-specific, overall, this process involves the removal of the plastic coating to the cable, when applicable, (Fig. 1a), installation of all the grommets on the cable spiral (sheath), threading of the inner cable, and mushrooming of the inner cable's remaining tip (Fig. 1b); the latter is important to ensure a strong connection between the cable and the injected terminal. Once the cables are assembled, they pass to the injection station where the second terminal is injected. This final process is performed in the production line BX726.

Three operations are performed at the injection stage: (1) injection of the cable terminal, (2) strength test, and (3) length measurement. In addition, two different terminal models are injected in the line BX726, shown in Fig. 2, these terminals are injected using the ZAMAK 5 alloy [21]. The cables are manually placed in the injection mould by an operator.

The measurement is automatically done with the cable taut, which is achieved by a 10 N load. If the cable length is correct, the cable is subject to an 800 N load. Then, the cable is transported by a robotic manipulator to another station where the sprues are removed by shearing. Regarding the ZAMAK injection process, it is performed within specific stations and is capable of around 600 injections/h. Amongst

Fig. 1 Preparation processes to the control cables: **a** removal of the plastic cover to the cable, **b** mushrooming of the inner cable's tip

Fig. 2 Types of cable ends injected in the line BX726: **a** model 75, **b** model 70

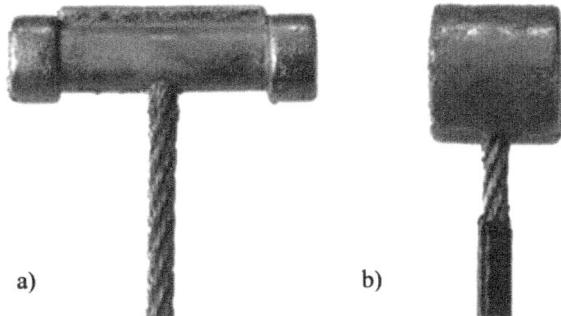

a)   b)

the subassemblies composing the injection machine, only the frontal support where the moulding takes is further studied in this work because of the moulds. The most used mould size in the company is 43 mm. The mould contains the terminals' final shape, cable guide grooves, and the sprue cavities; the latter being the object of this study.

2.2 Current Problematic

The main problem is the excessive size of the injection moulding sprues; in their current form, the sprues are non-uniform and big (Fig. 3), behaving randomly throughout the process. Consequently, their random breaking leaves them scattered around the facilities (Fig. 4). The large sprue size affects the final properties of the cable terminal because it alters the cooling rates, compromising the final quality of the product. In addition, the mean weight of a resulting sprue is 1.94 g/cable, and the line has a production capability of 450 cables/h; hence, large quantities of ZAMAK are scrapped (around 140.5 kg/month). Furthermore, in the current form, containers were placed in various parts as a palliative measure to collect the broken sprues with limited success, as shown in Fig. 4c, d.

2.3 Objectives

This work aims to reduce the sprue size on the control cables, while also having control over the sprues' size, allowing for its removal within a specific zone of the equipment. The necessary modifications are planned for the production line BX726, but it is in continuous operation; consequently, all the modifications should be done within a two-day period. In addition, the modifications should be as minimal as possible, allowing the interchangeability with other lines and reducing costs.

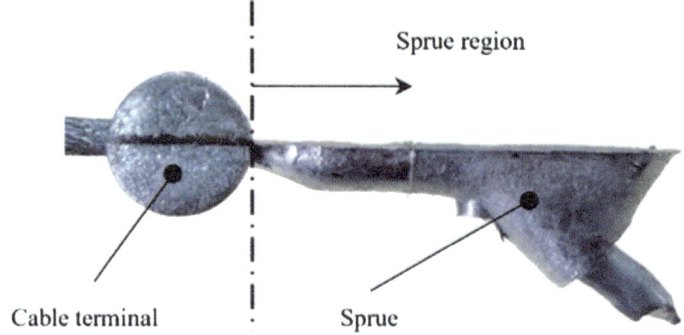

Fig. 3 Cable terminal with its sprue

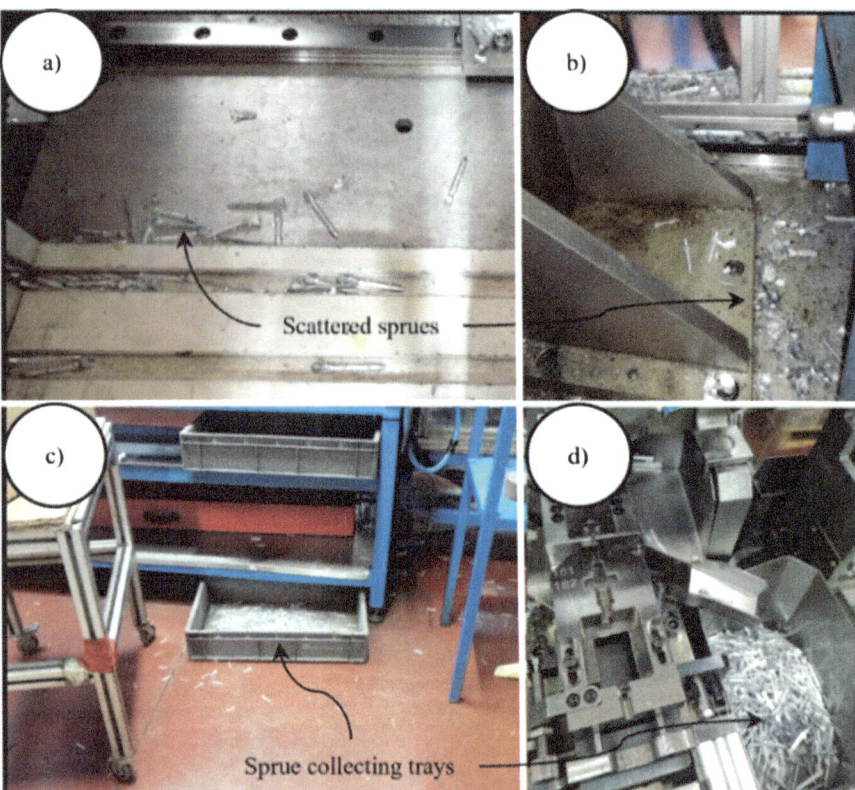

Fig. 4 Scattered sprues in the work floor and palliative measures to contain them

3 Methodology

3.1 Pre-design

The first stage consisted of the observation of the current process, leading to determining the possible modifications and improvements. Moreover, reducing the sprue size requires improvements to the mould structure. A common mould base eases the mould changing processes according to the production run. In this regard, several options were proposed and analysed using the SWOT method. The chosen design allows keeping the current mould bases, adapted now for 16 mm ones. This is suitable for 95% of the cable ends manufactured in the facilities. Furthermore, the mould was modified to inject cable terminals with sizes up to 7 mm. Finally, the final design was compact. Subsequently, modifications to the injection machine are necessary to accommodate the modified mould bases. The components to modify are the front and crucible supports. The modifications involved slight machining of four existing parts and the fabrication of one extra support. In addition, the modifications described above imply the displacement of the measuring kit by 22.4 mm forwards; thus, two current parts were slightly modified by machining whilst three auxiliary supports were fabricated. The modifications were found to fulfil the requirements without changing the actual working principle, remaining like the other machines in the production line. Moreover, the parts requiring fabrication were designed to be integrated with the current componentry.

The supporting structure of the robotic manipulator now interferes with the flow of control cables, thus requiring slight modifications. The modifications required refitment of the supports at one side of the structure (Fig. 5a) and were designed to be carried out on-site. An analysis using FEM was performed on the modified structure to ensure its performance under load. The assembly of the robotic manipulator on the structure and the new supports is shown in Fig. 5b. The structure was assumed to be built with steel C45E and was modelled as linear elastic.

a)   b)

Fig. 5 Modified structure for the robotic manipulator: **a** standalone structure, **b** assembled manipulator

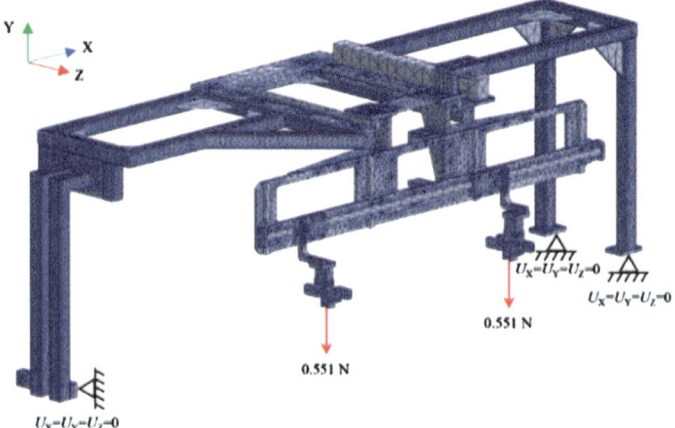

Fig. 6 Finite element mesh, boundary conditions and loads on the robotic manipulator's structure

Regarding the boundary conditions of these models, the bottom parts of the supports were considered fixed in all directions ($U_X = U_Y = U_Z = 0$). The load applied in the manipulator was 0.551 N, corresponding to the weight of the control cables. The imposed boundary conditions and loads are shown in Fig. 6. Afterwards, the structure was meshed with solid elements, four-node tetrahedrons, resulting in 203,484 elements and 404,962 nodes; the meshed structure is shown in Fig. 6. The analysis was solved as linear elastic under the assumption of large deformations.

The results from the FEM analyses indicated that the modifications do not affect the mechanical characteristics of the structure under working loads. In this case, the von Mises stress under working loads has a maximum value of 180 kPa whilst the yield strength of this material is 620 MPa, indicating that the applied load has no significant effect on the structure (Fig. 7). Similarly, the deformation of the structure was evaluated, being the displacement's magnitude of around 7 μm at the grips, which does not affect the function of the manipulator.

3.2 Workstation Interventions

The main subset of this intervention is the moulds and their structures due to the precision and heat treatments involved in the process. Subsequently, dimensional inspections were performed to validate their compliance with the design, followed by a functional validation, as described below.

Validation of the mould structure

The moulds were mounted in a testing injection machine within the company facilities, first, the original moulds were mounted and then all the necessary adjustments

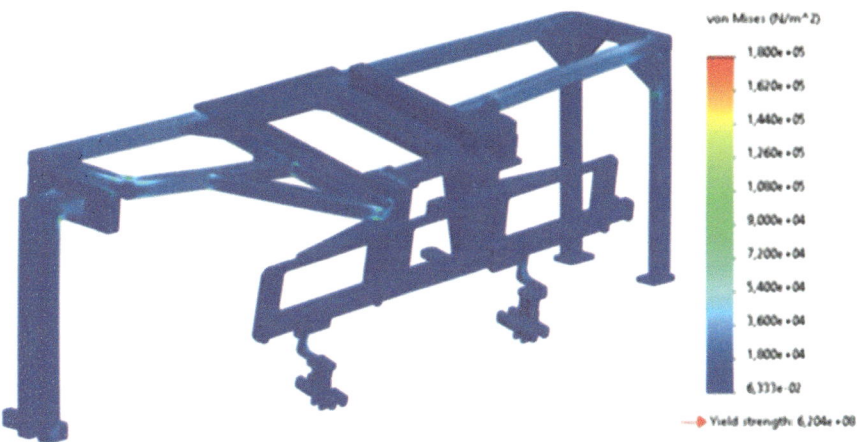

Fig. 7 Effective stresses in the robotic manipulator's structure under working loads

were done. The validation process consisted of 25 injections from each terminal model; then, the produced parts are inspected and measured, and the respective dimensional reports are performed. The cause of the premature sprue breaking was the original gating system section, which had reduced dimensions. The gating dimensions were adjusted using Electric Discharge Machining (EDM), as shown in Fig. 8. The mould structure also required adjustments to accommodate these changes. Subsequently, the second round of tests was performed, resulting in sturdier sprues which remained attached to the terminal throughout the process; therefore, the intervention was successful.

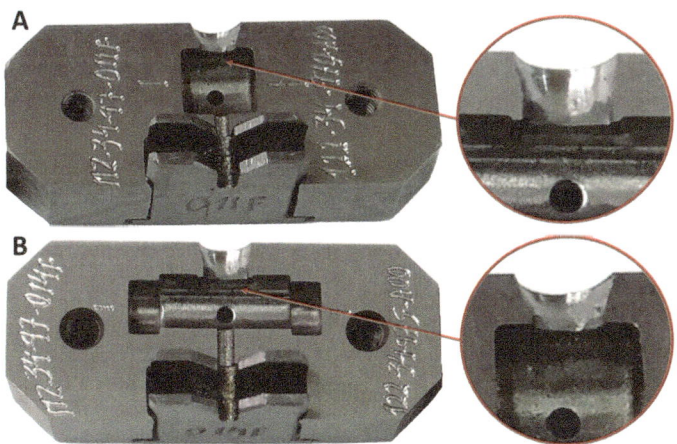

Fig. 8 Gating system, after modification, for terminal of model 75 (**a**) and model 70 (**b**)

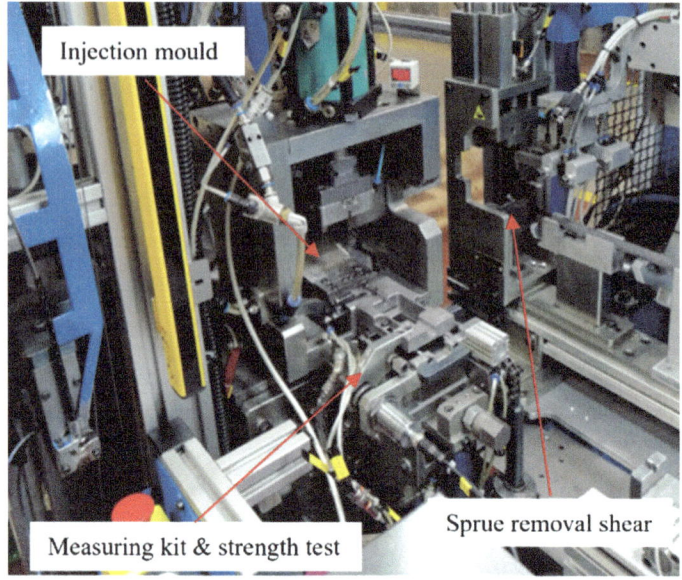

Fig. 9 Equipment after assembly and ready for operation

Machine interventions

All the components to be modified or replaced were removed from the machinery. Then, a specialized technician in ZAMAK injection machines was assigned to install, adjust, and validate the structure. The alteration to the structure of the robotic manipulator was performed in parallel to other tasks. Once the mentioned changes were implemented, all the conditions were met for the completion of the assembly of all components. Figure 9 presents the device with the changes completed.

To prepare the equipment for production, adjustments were made to the various subassemblies; subsequently, the final validation of the produced cables was performed. This step is performed on a measuring bench, concluding that the process fulfilled the requirements.

4 Results

4.1 Raw Material Waste

After completing the changes to the equipment, the first sprue samples were analysed and compared with those obtained before their implementation. The average mass difference from five samples of each batch was 0.93 g, being 48.3% lighter than the initial sprues. The reduction of 1.21 g (62.6%) achieved with this change in the mould

Fig. 10 Equipment after the intervention: correct sprue storage in the dedicated container (A), lack of scattered sprues (B), and clean path up to the container after continuous production (C)

structure, compared to the initial model, proved to be very positive for the economic analysis of the investment. In consequence, it is planned to implement these changes to 10 production lines in operation within the facilities. However, the cost analysis is not described here because it is out of the scope of this paper.

4.2 Sprue Control

Besides the sprue mass and respective cost savings with the proposed change, control of the process should be assured by guaranteeing the sprue collection in the dedicated container. To achieve this goal, changes were made to the feeding sections of the sprue removal shear, making it possible to reinforce them and avoid premature failure. With the accomplished changes and respective production monitoring, it was found that all sprues followed the injected terminals up to reaching the sprue removal station, and the existence of any sprues outside the container for their collection was eliminated, as shown in Fig. 10, contrasting with that previously shown in Fig. 4.

4.3 Final Product Quality Improvement

The moulds currently used by the company do not include exhaust ports for the gases, resulting in porosities within the parts. As the volume of the cavities dedicated to the sprue was reduced, a reduction in the porosities inside the injected terminals after the intervention is expected. To analyse the changes in the injection process, a few samples were prepared for a metallographic analysis by following the field's common practices and procedures; for porosity analysis, the BDG P202 standard [22] was followed. In addition, the images obtained with an Olympus BX53M Metallurgical Microscope and 200× optical zoom were processed in the Olympus

Stream Basic® software, making it possible to quantify the porosity within the terminals. As a consequence of the interventions, the porosity was reduced by 10.2 and 55.9% for terminal models 75 and 70. Nevertheless, this would be improved further by adjusting the mould lubrification systems. Furthermore, metallographic analyses were performed, the reduction of sprue size led to more controlled cooling, resulting in a finer grain size which has a positive effect on the mechanical performance of the injected terminal, e.g., higher fatigue strength [23].

5 Conclusions

In this work, the process of injecting the terminals to control cables was analysed and improved. Although the main problem was related to the sprues, it was necessary to find the cause of the problem and correct it, requiring modifications to several components within the production line.

The systematic analyses performed led to the improvement of the mould structures, mould sprue gates, measuring kit and strength testing device, the structure of the robotic manipulator, and auxiliary components of the injection moulding machine. In summary, the capability to inject 95% of the company products with one standardised type of mould structure led to the reduction of downtime. Furthermore, the number of fabricated components was minimal whilst existing components were modified slightly by machining; hence, maintaining compatibility with the other production lines operating in the company.

On the other hand, the random size of the sprues was caused by a variation on the sprue gate throughout the moulds used in the company. Consequently, moulds and mould structures were fine-tuned to produce a controlled sprue size, also reflected in the controlled behaviour of the sprues. The reduction of the sprue size led to the improvement of the terminal strength due to microstructure changes, which improved the quality of the product.

The combination of the SWOT and FEM techniques together with the synchronisation between departments within the company led to a significant improvement, which did not require major disruptions to the company's production. Furthermore, the investment for one production line will be returned in slightly over a year, whilst it would take four years for 10 production lines; consequently, the return of investment is relatively quick. In conclusion, the combination of process improvement techniques and engineering knowledge benefited the quality of a product, while also improving the reliability and reducing production costs of the associated machinery.

Declaration of Conflicting Interests The authors declare that there are no conflicts of interest regarding the publication of this paper.

References

1. Senthilkumar, C.B., Nallusamy, S.: Enrichment of quality rate and output level in a medium scale manufacturing industry by implementation of appropriate quality tools. Mater. Today Proc. **37**, 817–822 (2021). https://doi.org/10.1016/j.matpr.2020.05.832
2. Cheng, G., Li, L.: Joint optimization of production, quality control and maintenance for serial-parallel multistage production systems. Reliab. Eng. Syst. Saf. **204**, 107–146 (2020)
3. Ishikawa, K.: What Is Total Quality Control? The Japanese way. Prentice Hall, New Jersey, United States (1985)
4. Westcott, R.T., Duffy, G.L.: The Certified Quality Improvement Associate Handbook. ASQ Quality Press, Milwaukee, Wisconsin (2014)
5. Kogawa, A.C., Salgado, H.R.N.: Quality tools for a successful strategic management. Int. J. Bus. Process. Integr. Manag. **8**(3), 153–159 (2017). https://doi.org/10.1504/IJBPIM.2017.085394
6. Wilson, C.: Brainstorming and Beyond: A User-Centered Design Method. Morgan Kaufmann, Massachusetts, United States (2013)
7. Rodrigues, H., Silva, F.J.G., Morgado, L.G., Sá, J.C., Ferreira, L.P., Campilho, R.D.S.G.: A novel computer application for scrap reporting and data management in the manufacturing of components for the automotive industry. Procedia Manuf. **51**, 1319–1326 (2020). https://doi.org/10.1016/j.promfg.2020.10.184
8. Silva, F.J.G., Campilho, R.D.S.G., Ferreira, L.P., Pereira, M.T.: Establishing guidelines to improve the high-pressure die casting process of complex aesthetics parts. Transdiscipl. Eng. Methods Soc. Innov. Ind. **40**(7), 887–896 (2018)
9. Gürel, E., Tat, M.: SWOT analysis: a theoretical review. J. Int. Soc. Res. **10**(51), 994–1006 (2017)
10. Dyson, R.G.: Strategic development and SWOT analysis at the University of Warwick. Eur. J. Oper. Res. **152**(3), 631–640 (2004)
11. Karimi, M., Niknamfar, A.H., Niaki, S.T.A.: An application of fuzzy-logic and grey-relational ANP-based SWOT in the ceramic and tile industry. Knowl.-Based Syst. **163**, 581–594 (2019)
12. Goenka, M., Nihal, C., Ramanathan, R., Gupta, P., Parashar, A., Joel, J.: Automobile parts casting-methods and materials used: a review. Mater. Today: Proc. **22**, 2525–2531 (2020). https://doi.org/10.1016/j.matpr.2020.03.381
13. Ayar, M.S., Ayar, V.S., George, P.M.: Simulation and experimental validation for defect reduction in geometry varied aluminium plates casted using sand casting. Mater. Today: Proc. **27**, 1422–1430 (2020). https://doi.org/10.1016/j.matpr.2020.02.788
14. Groover, M.P.: Fundamentals of Modern Manufacturing: Materials, Processes, and Systems. Wiley, New York, United States (2020)
15. Anderson, B.: Die Casting Engineering: A Hydraulic, Thermal and Mechanical Process. CRC Press, Florida, United States (2005)
16. Pinto, H., Silva, F.J.G.: Optimisation of die casting process in Zamak alloys. Procedia Manuf. **11**, 517–525 (2017). https://doi.org/10.1016/j.promfg.2017.07.145
17. Pinto, H.A., Silva, F.J.G., Martinho, R.P., Campilho, R.D.S.G., Pinto, A.G.: Improvement and validation of Zamak die casting moulds. Procedia Manuf. **38**, 1547–1557 (2019). https://doi.org/10.1016/j.promfg.2020.01.131
18. Chandrasekaran, R., Campilho, R.D.S.G., Silva, F.J.G.: Reduction of scrap percentage of cast parts by optimizing the process parameters. Procedia Manuf. **38**, 1050–1057 (2019). https://doi.org/10.1016/j.promfg.2020.01.191
19. Nunes, V., Silva, F.J.G., Andrade, M.F., Alexandre, R., Baptista, A.P.M.: Increasing the lifespan of high-pressure die cast molds subjected to severe wear. Surf. Coat. Technol. **332**, 319–331 (2017). https://doi.org/10.1016/j.surfcoat.2017.05.098
20. Sun, J., Le, Q., Fu, L., Bai, J., Tretter, J., Herbold, K., et al.: Gas entrainment behavior of aluminum alloy engine crankcases during the low-pressure-die-casting process. J. Mater. Process. Technol. **266**, 274–282 (2019). https://doi.org/10.1016/j.jmatprotec.2018.11.016

21. ASM International: Zinc and Zinc Alloys (1998)
22. BDG: BDG Reference Sheet P202. BDG, p. 24 (2010)
23. Askeland, D.R.: The Science and Engineering of Materials. Springer, US, Boston, MA (1996)

Joining

Welding of High-Strength Steels for the Automotive Industry

T. Węgrzyn[ID], B. Szczucka-Lasota[ID], T. Szymczak[ID], B. Łazarz[ID], P. Cybulko[ID], and A. Jurek[ID]

Abstract Transport based on lighter and more durable vehicles is significant for different branches of industry. In automotive sector, high-strength martensitic steels play an essential role, because of yield stress and ultimate tensile strength. The article is focused on examining and comparing the behaviour of S960QL (Q—Quenching and Tempering, L—Low notch toughness testing temperature) steel as a representative of HSS grade steel and Docol 1200M (M—martensitic) steel as a representative of AHSS grade one. HSS and AHSS steels are relatively not very well weldable, as evidenced by the significant difference in the mechanical properties of the base material and the weld. The HAZ (Heat Affected Zone) and weld reflect cracks. Another serious disadvantage is related to the two times smaller value of ultimate tensile strength of a weld than in the case of base material. The differences in strengths are the consequence of chemical composition and microstructure of the base material and weld, which results as effect of electrode wires with increased nickel and molybdenum content and much lower sulphur and phosphorus content that affects much better mechanical properties. The weld metal deposit contains mainly martensite

T. Węgrzyn (✉) · B. Szczucka-Lasota · B. Łazarz
Silesian University of Technology, Krasinskiego 8, 40-019 Katowice, Poland
e-mail: tomasz.wegrzyn@polsl.pl

B. Szczucka-Lasota
e-mail: bozena.szczucka-lasota@polsl.pl

B. Łazarz
e-mail: boguslaw.lazarz@polsl.pl

T. Szymczak
Motor Transport Institute, Jagiellonska 80, 03-301 Warsaw, Poland
e-mail: tadeusz.szymczak@its.waw.pl

P. Cybulko
Institute of Power Engineering, Sw. Rocha 16, 15-879 Bialystok, Poland
e-mail: p.cybulko@iezd.com

A. Jurek
Novar Sp. z. o. o., Towarowa 2/8, 44-100 Gliwice, Poland
e-mail: adam.jurek@novar.pl

© The Author(s), under exclusive license to Springer Nature Switzerland AG 2023
L. F. M. da Silva et al. (eds.), *1st International Conference on Engineering Manufacture 2022*, Proceedings in Engineering Mechanics,
https://doi.org/10.1007/978-3-031-13234-6_6

and bainite with coarse ferrite, while the parent material contains mainly martensite and fine ferrite. Non-destructive inspections on macro-specimens were corresponded with further destructive test results (tensile strength, hardness, fatigue, metallographic structure analyses). Main aim of the article is to select one of the tested materials for the construction of a movable landing, taking into account the properties of the joint after welding with the use of various parameters.

Keywords Welding · Properties · High-Strength Steel · Characterization · Static · Fatigue

1   Introduction

The engineering and scientific groups at new material approaches to structures are focused on applications of high-strength steels, because they enable obtaining a mass reduction of various types of construction, collecting bodies in white and vehicles, finally [1–3]. Important material for this type of application is high-strength steel (HSS): S960QL (Q—Quenching and Tempering, L—Low notch toughness testing temperature) steel and advanced high-strength steel (AHSS steels): DOCOL 1200M (M-martensitic). The ultimate tensile strength of HSS grade steel usually reaches 1000 MPa, while in the case of AHSS grade steel, this parameter exceeds the value indicated. Both tested steel grades are considered relatively difficult to weld due to the significant differences in the microstructure [4–7]. After welding, cracks often occur in joints covering the weld and the Heat Affected Zone (HAZ).

The welding technology is complicated because pre-heating and interpass temperatures are required. Depending on the thickness of the sheets used for components, it is essential to chamfer the edges of the sheets and carefully select electrode wires, protective gases and voltage protection parameters [8–10].

The article presents the welding problem for the grade steels in two different manufacturing states, i.e. Q—Quenching and Tempering (S960 QL) and Hardening (DOCOL 1200M). It was decided to check the weldability of these steels in terms of their application in the thin-walled construction having 6 mm thickness concerning its application in the mobile platform.

2   Materials and Methods

High-strength steel S960QL (recommended for crane components) and advanced high-strength steel DOCOL 1200M (M-martensitic) steel (used for car safety cage) were selected. During welding of these steel grades, a reduction of mechanical properties in HAZ can be observed. Therefore, it is recommended to limit the linear energy during welding to the level of 5 kJ/cm [2, 3, 11, 12]. Figure 1 presents the

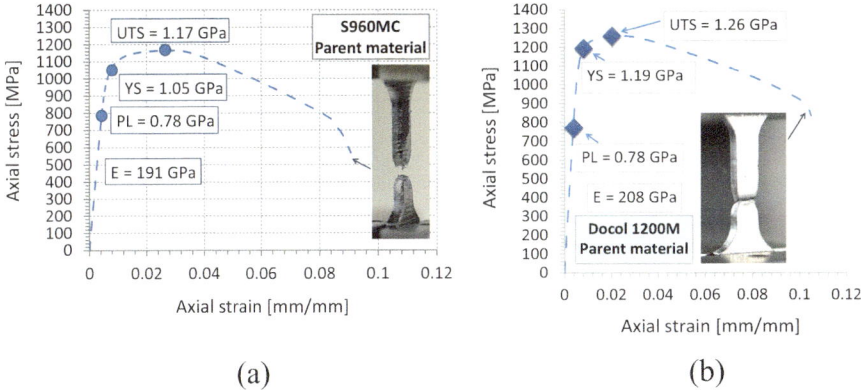

Fig. 1 Tensile characteristics and mechanical parameters of: S960 QL (**a**) and DOCOL 1200M (**b**) steels, E—Young's modulus, PL—proportional limit, YS—yield stress, UTS—ultimate tensile strength [13, 14]

mechanical properties of the tested steel grades, obtained by authors in previous publications [13, 14].

The high strength of both types of steel results from the chemical composition, which is significantly different in relation to low-alloy steels [12, 15] (Table 1).

The high value of ultimate tensile strength of the material is related to the higher content of carbon, vanadium, niobium and especially titanium. In C-Mn steels, the content of Ti is introduced at the maximum level of 0.002%, and in high-strength steels, the content of titanium is ten times higher (Table 1).

It was decided to make a joint of both steel grades with a thickness of 6 mm using the MAG process at two different Ar-CO_2 shielding mixtures as recommended [2, 3, 11]. Additionally, two UNION X96 electrode wires (EN ISO 16834-A G 89 6 M21 Mn4Ni2CrMo) and the UNION X90 wire (EN ISO 16834-A G 89 6 M21 Mn4Ni2CrMo) with the following chemical composition (Table 2) were tested.

Table 1 Chemical composition of S960 QL and DOCOL 1200M steels [13, 14]

Meterial	C (%)	Si (%)	Mn (%)	P (%)	S (%)	Al (%)	Nb (%)	V (%)	Ti (%)
S960QL	0.12	0.25	1.3	0.02	0.01	0.015	0.05	0.05	0.017
DOCOL 1200M	0.14	0.21	1.3	0.008	0.001	0.045	0.015	0.01	0.025

Table 2 Chemical composition of electrodes wires [13, 14]

UNION	C (%)	Si (%)	Mn (%)	P (%)	Cr (%)	Mo (%)	Ni (%)	Ti (%)
X90	0.10	0.8	1.8	0.010	0.35	0.6	2.3	0.005
X96	0.11	0.8	1.8	0.010	0.45	0.65	2.45	0.007

The chemical composition of both wires (Table 2) is different from the chemical composition of the welded steels (Table 1). The addition of chromium in the electrode wire (which is not an alloying element of S960QL steel or DOCOL 1200M steel) is also essential. It was decided to select the welding parameters for both steel grades employing two various electrode wires.

The welding parameters were as follows: the diameter of the electrode wire was 1.0 mm, the arc voltage was 29.5 V, and the welding current was 123 A. The weld was of a multi-run. The welding speed has reached 430 mm/min. The shielding mixture in the MAG process was varied twice:

- Ar + 15% CO_2,
- Ar + 25% CO_2.

The joints were manufactured with pre-heating to 115 °C and without pre-heating. The interpass temperature was not controlled.

After MAG welding using two various mixtures (Ar + 15% CO_2 and Ar + 25% CO_2) and two tested electrode wires (UNION X90 and UNION X96), non-destructive and destructive tests were carried out. As part of non-destructive testing (NDT), the following ones were conducted:

- visual testing (VT) at magnification 3×. The inspections were led to the requirements of the PN-EN ISO 17638 standard. Fulfil criteria have followed EN ISO 5817 standard,
- magnetic particle testing (MT)—the tests were carried out following the PN-EN ISO 17638 and EN ISO 5817 standards, with a magnetic flaw detector test device type REM—230.

Three-point bending tests were conducted at room temperature using cuboid specimens having a thickness of a = 6 mm and width b = 20 mm. The mandrel diameter was equal to d = 22 mm, at the distance between its supports: d + 3a = 31 mm. The required angle of the bending test was represented by a value of 180°. Five bending test measurements were carried out for each tested joint thickness on the root side and the face side.

The final stage of the experimental procedure was represented by the Charpy impact test at two values of temperature, i.e. 0 and −20 °C.

3   Results

The welds without pre-heating were noticed as not good joints, because of cracks occurring. Therefore, it was decided to pre-heat all tests to a temperature of 115 °C.

The results of non-destructive examining of the tested joints (with pre-heating) are shown in Table 3.

It was noted that the influence of the shielding gas mixture is much more important than the choice of the electrode wire. It means gas mixture Ar + 15% CO_2 could be treated as a better choice. The Ar + 25% CO_2 mixture has a more oxidizing nature

Table 3 Result of non-destructive testing of a joint

Material	Shield gas	Wire electrode	Observation
S960QL	Ar + 15% CO_2	UNION X90	No cracks
S960QL	Ar + 15% O_2	UNION X96	No cracks
S960QL	Ar + 25% CO_2	UNION X90	Cracks in the HAZ
S960QL	Ar + 25% CO_2	UNION X96	Cracks in the HAZ and weld
DOCOL 1200M	Ar + 15% CO_2	UNION X90	No cracks
DOCOL 1200M	Ar + 15% O_2	UNION X96	No cracks
DOCOL 1200M	Ar + 25% CO_2	UNION X90	Cracks in the HAZ
DOCOL 1200M	Ar + 25% CO_2	UNION X96	Cracks in the HAZ and weld

of the weld metal than the Ar + 15% CO_2 mixture, which promotes the formation of cracks in the weld. In the more oxidized weld metal, there are usually higher amounts of oxide inclusions having a direct influence on the initiations of cracks in the joint.

For further (destructive) tests, the joints made with two tested electrode wires were taken into account, but only using a shield gas mixture of the Ar + 15% CO_2 with pre-heating to a temperature of 115 °C. Results of tensile tests (average of 3 tests) are presented in Table 4. The table data show that a high value of ultimate strength (at the level of 700 MPa) and acceptable relative elongation (at 8%) were obtained only at DOCOL 1200M welding.

The joints made for S960QL steel were not at a pretty good level, but slightly worse. It was also decided to weld the mechanical properties of both materials using a consistently shielding mixture the shielding argon gas mixture containing 15% CO_2.

The test results are summarized in Table 5.

No cracks were found in all examined cases (Table 5). These results prove that the joint has good plastic properties. This was also confirmed in a microstructure

Table 4 The results of tensile tests

Material	Shielding gas	Wire electrode	UTS (MPa)	A_5 (%)
S960QL	Ar + 15% CO_2	UNION X90	603	7.8
S960QL	Ar + 15% CO_2	UNION X96	615	7.9
DOCOL 1200M	Ar + 15% CO_2	UNION X90	721	8.2
DOCOL 1200M	Ar + 15% CO_2	UNION X96	723	8.3

Table 5 The results of bending tests

Material	Wire	Deformed side	Notes
S960QL	UNION X90	Root of weld	No cracks
S960QL	UNION X90	Face of weld	No cracks
S960QL	UNION X96	Root of weld	No cracks
S960QL	UNION X96	Face of weld	No cracks
DOCOL 1200M	UNION X90	Root of weld	No cracks
DOCOL 1200M	UNION X90	Face of weld	No cracks
DOCOL 1200M	UNION X96	Root of weld	No cracks
DOCOL 1200M	UNION X96	Face of weld	No cracks

analysis by the occurrence of three structure components: martensite, bainit and ferrite (Fig. 2).

Generally, martensite, bainite and ferrite were observed in both welds. Ferrite in DOCOL 1200M was clearly more fragmented. This can be explained by the increased content of Ti in the DOCOL 1200M than in S960QL, which may translate into the occurrence of TiO and TiN inclusions on which fine-grained ferrite is easily nucleated.

Table 6 data shows that after S960QL welding, the criterion of 47 J at 0 °C is fulfilled only in one case, i.e. after the use of the UNION X90 electrode wire and other recommendations resulting from the previous part of the tests:

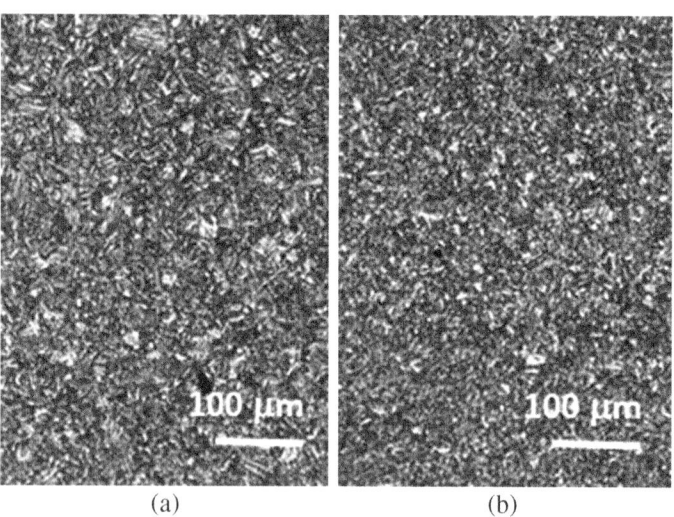

(a)   (b)

Fig. 2 Microstructure of the weld (**a** S960QL, **b** DOCOL 1200M). Martensite is the dominant phase in both cases

Table 6 The results of Charpy impact tests

Material	Wire	KV (-20 °C)	KV (0 °C)
S960QL	UNION X90	33	53
S960QL	UNION X96	27	41
DOCOL 1200M	UNION X90	48	65
DOCOL 1200M	UNION X96	41	55

- pre-heating temperature at 110 °C,
- use of a shielding gas mixture of Ar + 15% CO_2. The table data also indicates that after DOCOL 1200M welding, the criterion of 47 can also be also preserved at -20 °C. This means that the impact resistance after welding DOCOL 1200M steel is much better than that 960 QL steel.

The welded joints' behaviour under monotonic tension was followed based on data for the parent materials (Fig. 1), comparison of the stress–strain curves (Fig. 3) and values of mechanical parameters, Table 7. For this case, a significant lowering of the mechanical parameters concerning welding technology was observed, reaching 50% in comparison to the material in the as-received state. In contrast, an increase of the relative elongation value was observed. Comparing these data with results for parent material (Fig. 3a), which illustrate significant differences in the tested steel responses, has enabled us to observe reduction of this feature, Fig. 3b. This allows us to formulate the following sentence: in the case of the S960MC and DOCOL 1200M steel MAG welding process led to diminishing differences in the material response under monotonic tension.

Results from Fig. 3 also indicate differences in the MAG welds behaviour under tensile cycles would be not observed. Nevertheless, this sentence is only an interpretation, and it should be checked experimentally. For this case, the fatigue test should

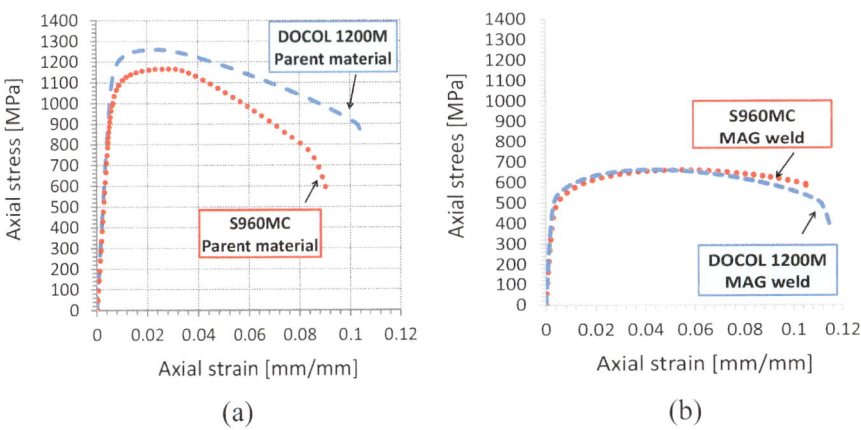

(a)   (b)

Fig. 3 Comparison of the tensile characteristics of the S960MC steel and DOCOL 1200M: **a** parent material, **b** MAG weld

Table 7 Mechanical parameters of parent materials and welded joints of S960MC and DOCOL 1200M

Mechanical properties of parent material and its MAG weld				
Type of material	Young's modulus (MPa)	Proportional limit (MPa)	Yield stress (MPa)	Ultimate tensile strength (MPa)
Parent material				
S960MC	191,204	778	1050	1166
DOCOL 1200M	207,858	771	1188	1259
Weld				
S960MC	199,340	185	538	661
DOCOL 1200M	220,919	274	533	663

be designed at a positive value of stress ratio (R), and therefore, its value equal to 0.1 was used. Based on data from the experiment under cyclic loading (Fig. 4), variations in the weld joint behaviour can be evidenced even at almost negligible differences in the static characteristics are noticed, Fig. 3b. As it was investigated, values of the fatigue limit of S960MC and DOCOL 1200M welded joints were varied by 4.2, Fig. 4. This kind of data can be directly used for the determination of proportion between the value of fatigue limit and ultimate tensile strength, extending knowledge on the MAG welded joint response for engineers and scientists with respect to inspections and modelling as well. This quotient was represented by the following values: 0.75 and 0.18 for the DOCOL 1200M and S960MC, respectively. With respect to engineering practice, the same approach can be used for proportional limit, which is related to material behaviour at an elastic state. This proportion has reached the values expressed by 0.63 and 1.81 for the S960MC and DOCOL 1200M, respectively.

Fig. 4 Wöhler curves of S960MC and DOCOL 1200M welded by means of MAG technique at UNION X96 welding wire supported, determined at stress ratio R = 0.1

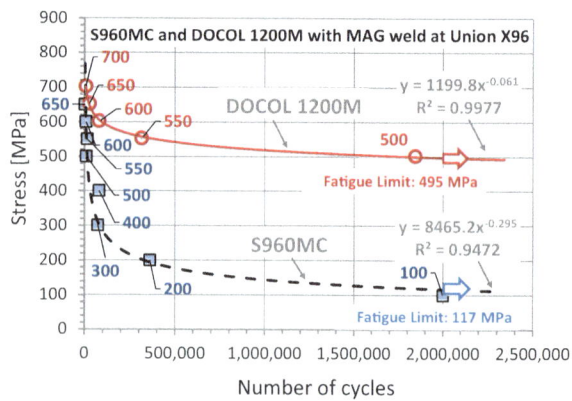

4 Summary

In the article, it was decided to carefully examine the possibilities of obtaining the correct joint made of S960QL and DOCOL 1200M steel. In the first part of the study, it was determined that it is essential to use pre-heating at the level of 115 °C. NDT showed that a less oxidizing gas mixture (15% carbon dioxide) enables obtaining joints without welding defects and incompatibilities. For this case, destructive tests, i.e., tensile, bending and Charpy impact tests follow the response of a joint.

The DOCOL 1200M has reflected a lower sensitivity on the MAG welding process than the S960MC steel.

The DOCOL 1200M welded joint has expressed a higher resistance on cyclic loading than 960QL steel.

Based on all the tests, the following conclusions can be proposed:

- pre-heating should be used,
- the shielding gas mixture of Ar-15% CO_2 is preferable to Ar-25 CO_2 due to its less oxidizing nature,
- UNION X90 electrode wire is preferred over UNION X96 wire due to its lower carbon content.

Acknowledgements The article is related to the implementation of the COST project, CA 18223.

References

1. Jaewson, L., Kamran, A., Jwo, P.: Modeling of failure mode of laser welds in lap-shear speciments of HSLA steel sheets. Eng. Fract. Mech. **78**(2), 374–396 (2011). https://doi.org/10.1016/j.engfracmech.2010.10.011
2. Prijanovič, U.: Marica Prijanovič Tonkovič, Uroš Trdan, Matej Pleterski, Matija Jezeršek Damjan Klobčar Remote fibre laser welding of advanced high strength martensitic steel. Metals **10**(4), 533 (2020). https://doi.org/10.3390/met10040533H
3. Lahtinen, T., Vilaça, P., Peura, P., Mehtonen, S.: MAG welding tests of modern high strength steels with minimum yield strength of 700 MPa. Appl. Sci. **9**, 1031 (2019). https://doi.org/10.3390/app9051031
4. Tomków, J., Haras, J.: The influence of welding heat input on the quality and properties of high strength low-alloy dissimilar steel butt joints. Weld. Tech. Rev. **92**(2), 15–23 (2020). https://doi.org/10.26628/wtr.v92i2.1091
5. Mert, T., Tümer, M., Kerimak, Z.M.: Investigations on mechanical strength and microstructure of multi-pass welded S690QL HSLA steel using MAG and FCAW. Pract. Metallogr. **56**(10), 634–654 (2019). https://doi.org/10.3139/147.110578
6. Sága, M., Blatnická, M., Blatnický, M., Dižo, M., Gerlici, J.: Research of the fatigue life of welded joints of high strength steel S960 QL created using laser and electron beams. Materials **13**, 2539 (2020). https://doi.org/10.3390/ma13112539
7. Totten, G., Howes, M., Inoue, T.: Handbook of Residual Stress and Deformation of Steel, p. 499. ASM International, Ohio, OH, USA (2002)
8. Hadryś, D.: Impact load of welds after micro-jet cooling. Arch. Metall. Mater. **60**(4), 2525–2528 (2015). https://doi.org/10.1515/amm-2015-0409

9. Celin, R., Burja, J.: Effect of cooling rates on the weld heat affected zone coarse grain microstructure. Metall. Mater. Eng. **24**(1), 37–44 (2018). https://doi.org/10.30544/342

10. Walsh, S.M., Smith, J.P., Browne, E.A., Hennighausen, T.W., Malouin, B.A.: Practical concerns for adoption of microjet cooling. In: ASME Proceedings 2018 Power Electronics, Energy Conversion, and Storage. https://doi.org/10.1115/IPACK2018-8468

11. Hadryś, D., Węgrzyn, T., Piwnik, J., Wszołek, L., Węgrzyn, D.: Compressive strength of steel frames after welding with micro-jet cooling. Arch. Metall. Mater. **61**(1), 123–126 (2016). https://doi.org/10.1515/amm-2016-0023

12. Skowrońska, B., Szulc, J., Chmielewski, T., Golański, D.: Wybrane właściwości złączy spawanych stali S700 MC wykonanych metodą hybrydową plazma + MAG. Weld. Technol. Rev. **89**(10), 104–111 (2017). https://doi.org/10.26628/ps.v89i10.825

13. Szymczak, T., Szczucka-Lasota, B., Węgrzyn, T., Łazarz, B., Jurek, A.: Behavior of weld to S960MC high strength steel from joining process at micro-jet cooling with critical parameters under static and fatigue loading. Materials **14**(11) (2021). https://doi.org/10.3390/ma14112707

14. Szczucka-Lasota, B., Węgrzyn, T., Szymczak, T., Jurek, A.: High martensitic steel after welding with micro-jet cooling in microstructural and mechanical investigations. Materials **14**(4) 936, 1–16 (2021). https://doi.org/10.3390/ma14040936.

15. Szczucka-Lasota, B., Gajdzik, B., Wegrzyn, T., Wszołek, Ł.: Steel weld metal deposit measured properties after immediate micro-jet cooling. Metals **7**(9), 339, 1–9 (2017). https://doi.org/10.3390/met7090339

Evaluation of Different Routes for Manufacturing of Micro Process Devices

T. Gietzelt[ID]**, V. Toth, T. Wunsch, and M. Kraut**[ID]

Abstract Micro process devices offer a high surface-to volume ratio, e.g. facilitating high amounts of heat transfer. By this, strongly exothermic chemical reactions may be transferred from batch to continuous processes, reaching higher product yield and decreasing by-products. For manufacturing of micro process devices, microstructuring and joining of the apparatuses accounts for the overwhelming part of the cost. Various technologies can be used for microstructuring. They have certain constraints, advantages and disadvantages, which should be considered by the designer in advance. Due to small wall thickness between reaction and cooling passages, and for reasons of strength and corrosion resistance, additional materials, such as those common in brazing, should be avoided. Thus, mainly diffusion bonding and laser welding are applicable for joining. Again, these technologies has certain constraints, advantages and disadvantages, which may interact with microstructuring technologies. Experience has shown that the selection of suitable methods by the designer determines success or failure of manufacturing of micro process devices. For this reason, in this publication special features of the above-mentioned technologies and dependencies are discussed and illustrated by several practical examples.

1 Introduction

Micro process technology offers great opportunities for process intensification. By conversion of raw materials more efficiently, higher yields are achieved and waste product flow is reduced. Due to the high heat transfer capacities of microstructured devices, batch processes can be converted into continuous processes. The volumes handled can be kept low, improving inherent safety.

However, in the transition from the macroscopic world to micro-process engineering, other aspects become more relevant, which may cause failures. Due to

T. Gietzelt (✉) · V. Toth · T. Wunsch · M. Kraut
KIT, Institute for Micro Process Engineering, 3640, 76021 Karlsruhe, Germany
e-mail: thomas.gietzelt@kit.edu

© The Author(s), under exclusive license to Springer Nature Switzerland AG 2023
L. F. M. da Silva et al. (eds.), *1st International Conference on Engineering Manufacture 2022*, Proceedings in Engineering Mechanics,
https://doi.org/10.1007/978-3-031-13234-6_7

small channel cross sections, pressure losses and flow rate variations become much more critical.

The impact of surface roughness increases for microchannels. Other issues are blocking of microchanels by impurities or plaques from side reactions or fouling.

Information on the corrosion rates of materials are no longer helpful, since the wall thicknesses between different passages are less than 1 mm. In addition, concentrations and temperatures can deviate significantly from those known from macroscopic apparatuses due to more intensive mixing and increased heat tone.

Therefore, for apparatuses that have high requirements for corrosion resistance, additional materials as common for brazing, should be avoided. Instead, technologies like diffusion bonding or laser welding are applied. By diffusion bonding, internal cross sections can be joint. By grain growth across bonding planes, monolithic components with superior pressure resistance are obtained. However, since the process requires very high temperatures of about 80% of the melting temperature and long dwell times, some materials are subjected to undesired changes, and corrosion resistance is decreased.

In laser welding, the weld seam geometries and seam cross sections can be tailored within a wide range. However, microchannels can lead to unexpected phenomena in terms of heat dissipation. In contrast to diffusion bonding, for laser welding the heat-affected zone remains locally restricted, and material properties unaffected.

Various technologies are available for microstructuring. Each has specific advantages and disadvantages as well as limitations in terms of freedom of design.

Interactions between microstructuring methods and joining processes must also be considered.

The task of the designer is to anticipate the complete manufacturing process and to consider the possibilities and limitations in advance. This may sound trivial or self-evident, but experience shows that this is often the reason for failure.

In this publication, the above-mentioned aspects are presented in detail, discussed and illustrated with selected case studies.

2 Technologies for Manufacturing of Microstructure

2.1 Mechanical Methods

All mechanical structuring methods suffer from serial work flow and low removal rate of material. It becomes worse for decreasing tool diameter, since stability of tools drop, and feed must be reduced.

All technologies of mechanical microstructuring involve chipping of material. Depending on the size and microscopic geometry of the cutting edges, interactions with the work piece occurs, e.g. compression, depending on the materials cutting behavior which may be brittle, ductile or tough. Therefore, tendency to cold work hardening or burr formation may arise. This may cause a number of problems, e.g.

distortion of thin sheet by bending and subsequently problems during stacking of multiple sheet assemblies. Burr formation may deteriorate vacuum tightness after diffusion bonding or to fill subsequently microstructures with catalyst [1].

Removal of burrs is not practicable in most cases for microstructured sheets, either due to the pure effort, damaging of the surface by scratching or local removing surplus material, causing tolerance issues of stacks consisting of multiple sheets, smearing of passivation layers and so on.

Hence, burr formation should be avoided if possible by adjusting appropriate machining parameters and quality control. This needs skilled staff and/or preliminary tests, depending on the sort of material to be machined.

2.1.1 Endmilling

End mills made of Hard Metal

End mills made of hard metals can be used for plastics, all sort of metals and even for some sorts of presintered or other machinable ceramics (e.g. glassy ceramic [MACOR™]).

Hard metals are composite materials, consisting of a soft binder matrix and embedded wear resistant particles, predominantly tungsten carbide. In consequence, shaping by grinding cause notchy cutting edges due to different material resistance to abrasion (Fig. 1).

As binder metals, mostly cobalt is used, but also nickel and iron is employed. Its content varies from 6 to 30%, depending on striking or continuous interaction of tool and work piece (turning $\geq$ percussion drilling). Since the wear is born by the carbide particles only, hard metals with low binder content are more wear resistant. In return, higher binder content improves flexural strength. Also decreasing carbide particle size is favorable in terms of flexural strength. Nowadays, fine-grained hard

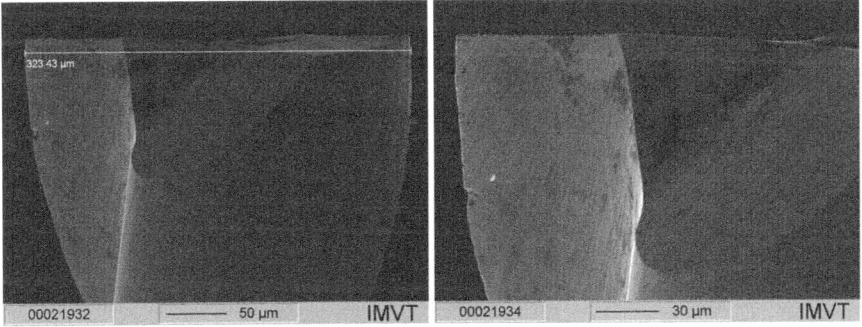

Fig. 1 Left: Uncoated end mill (HAM, Ø = 0.3 mm) and detail of micro-notched cutting edge (right)

metals with carbide particle sizes down to 0.2 μm are available, possessing superior flexural strength and good wear resistance [2–5].

Recently, end mills with two cutting edges and made of ultrafine-grained hard metal are available down to diameters of 30 μm [6]. From a logical point of view, the mechanical properties become more and more inhomogeneous, if the tool diameter decreases to the range of the hard metals microstructure. It has to be pointed out, that the manufacturing yield of end mills obviously drops with diameter, since prices increase. Also durability and lifetime scatters considerably (Fig. 2).

To improve wear resistance, most tools are coated with additional hard layers deposited by PVD (Physical Vapor Deposition) [7–10]. Thickness, quality and adhesive strength of these layers may vary from batch to batch. Besides better wear resistance, coatings increase rounding of the cutting edge, increasing machining forces, and facilitates burr formation (Figs. 3 and 4).

Several adapted shapes of end mills are available, and the quality and price varies for different suppliers [11, 12]. For manufacturing of deep and narrow trenches, end mills possessing a small length with two flutes at the tip only, but circular cross section above and conicity at a certain angle, to improve resistance to bending, were implemented (Fig. 5). Increasing the number of cutting edges to more than two, cause

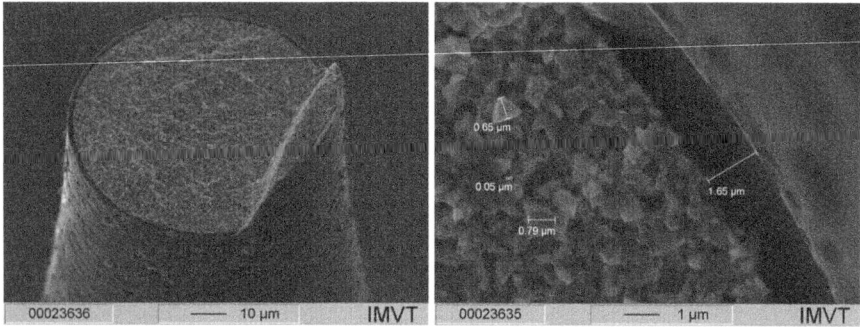

Fig. 2 Cross section of a broken ultrafine-grained hard metal end mill (∅ = 100 μm). PVD-coating thickness app. 1.6 μm

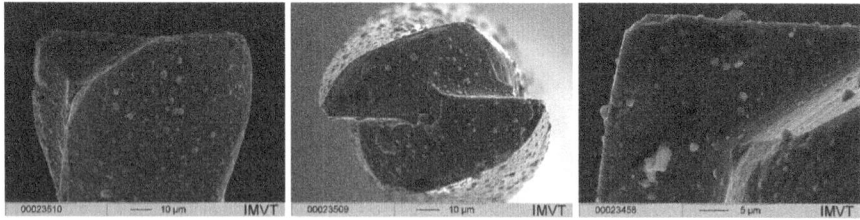

Fig. 3 PVD-coated 100 μm-end mill with droplet formation and flaking of coating. Right: Rounding of cutting edge by coating

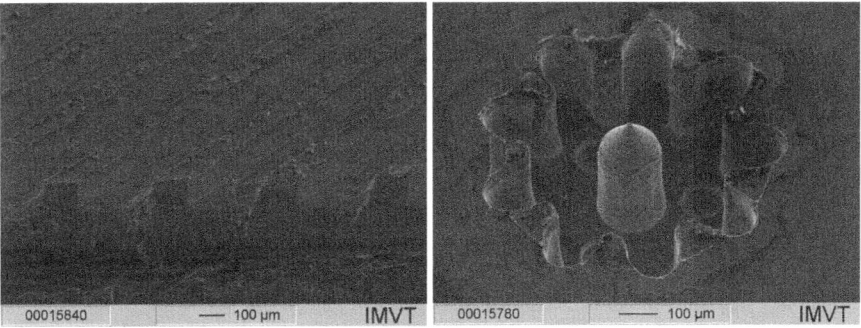

Fig. 4 Examples for burr formation at mechanically milled microstructures. Left: Milled trench. Right: Micro gearwheel mold insert

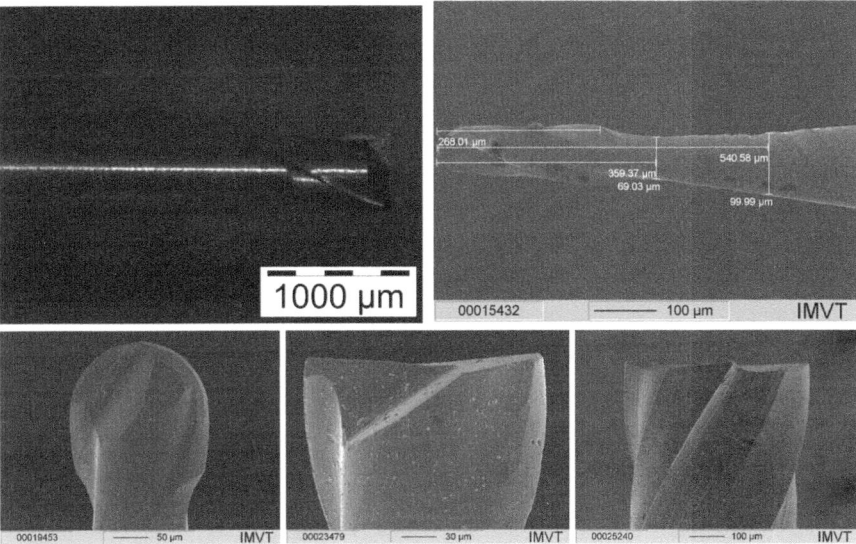

Fig. 5 Different shapes of hard metal end mills. Top left: fluted tip (GDE, $\emptyset = 0.6$ mm), top right: conical shaft (SPPW, $\emptyset = 0.1$ mm). Bottom left: radius end mill (NS tool, $\emptyset = 0.2$ mm). Middle: End mill without radii (Hitachi, $\emptyset = 0.2$ mm). Right: End mill without radii and four flutes (Mitsubishi, $\emptyset = 0.5$ mm)

decreasing chipping space but improve stability and productivity if rotational speed and feed can be increased accordingly.

For very small features and milling tools possessing a long cylindrical shaft for large machining depths, chip removal can be problematic.

In distinction to macroscopic machining, for decreasing end mill diameters very high rotational speed is required to achieve reasonable cutting speed. High speed

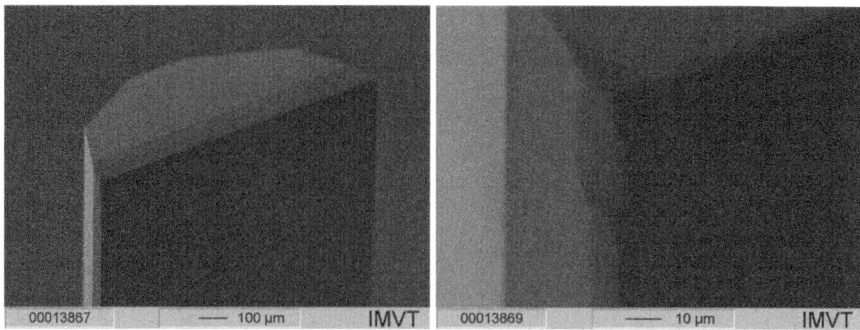

Fig. 6 Tool made of monocrystalline diamond and detail of cutting edge with wear mark

spindles up to 150.000 rpm are common, but 10.000 rpm should be exceeded at least.

Accordingly, the feed per cutting tooth determines tool load. To prevent squeezing and guaranteeing chipping, a minimum feed of 1 μm per cutting edge should be ensured. The infeed should be kept low due to limited stability of small diameter end mills.

As a consequence, eccentricity of tool clamping becomes important. Mechanical collets are worse compared to thermal shrinking chucks.

The machine dynamics and processing speed of the NC control must ensure to achieve high feed rates at complex outlines.

End mills made of Monocrystalline Diamond

Tools made of monocrystalline diamond can be used for nonferrous materials like copper and brass only, due to heat generation during chipping and excessive diffusivity of carbon in iron, titanium and others. Tools possess only one cutting edge which is ideally sharp. The quality of the cutting edge helps to prevent burr formation. Costs are high and care must be taken to prevent tool fracture. Only a few suppliers are available, mostly with jewelry background and knowledge how to shape diamonds [13, 14]. Figure 6 displays a diamond tool with a detailed view of a wear mark at the cutting edge after excessive use.

2.1.2 Sawing and Flycutting

For fly cutting, a cutting edge is placed at the perimeter of a rotational tool. As for sawing, a lead-out area is required around the structure to be manufactured according to the diameter of the tool. The shape of the cutting edge corresponds to the microchannel cross-section, and only straight structures can be produced [15].

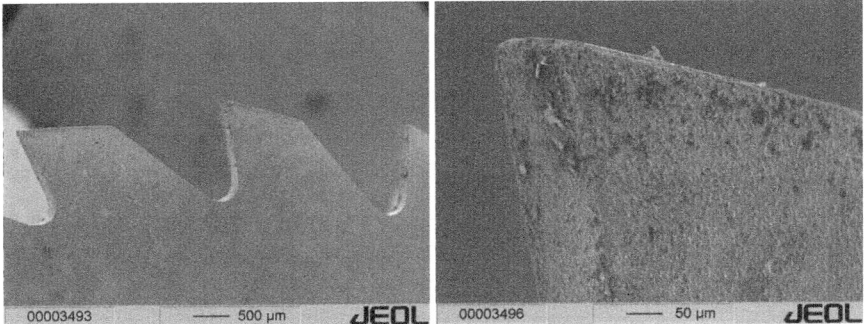

Fig. 7 Saw blade with 30 teeth and wear resistant PVD-coating

Due to multiple cutting edges for saw blades, productivity may be high, especially, if multiple sawing blades are fixed at one spindle (Fig. 7). However, due to the standardized shape of saw blades, only rectangular cross sections for microchannels can be manufactured.

2.1.3 Slotting

Slotting tools can be grinded from old end mills according to the microstructures to be manufactured. Different cross sections of microchannels are possible. The advantage compared to end mills is a much better stability, especially for small trench width (Fig. 8). Normally, only straight microchannels can be slotted. However, with restrictions, some curved trenches according the shape of the tool, are possible. To do so, the tool must be mounted on a rotary axis and carefully adjusted to the center. Then it can be turned with respect to the current direction of travel of the workbench.

The productivity or material removal rate, however, is very slow.

2.2 Chemical Etching

Chemical etching is offered as a service by several companies for a wide range of materials [16–20]. Unfortunately, not all information on etchable materials provided by the suppliers match claims: Tantalum and nickel base alloys are not possible to etch. Mainly iron-based alloys are etched using a $FeCl_3$ process.

The depth-to-width ratio is limited to 0.8. Perpendicular walls cannot be manufactured. Instead, ellipsoid channel cross sections are formed, facilitating fluidic flow. The surface roughness is about one order of magnitude higher than for milled microchannels, increasing pressure loss and facilitating fouling.

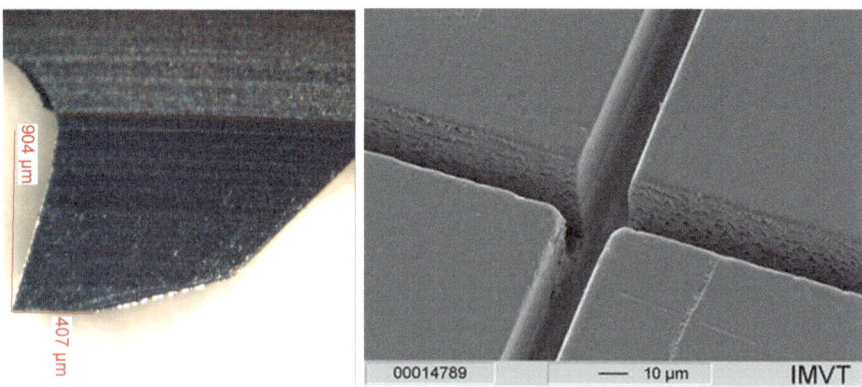

Fig. 8 Left: Side-view of a grinded slotting tool. Right: Slotted trench (app. 15 μm wide, 150 μm deep, PMMA)

Depending on the remaining residual bottom thickness and the homogeneity of the material, for micro process devices leakage between different passages may arise due to pores and nonmetallic inclusions.

When etching is performed from both sides, alignment problems of the photomask may arise for very small features, since both sides must be exposed with the same mask. In the middle of sheet thickness, where microstructures from both sides meet, an etched fin is formed (Fig. 9). Also uniformity of etching across large areas may be a problem [21].

Advantages are freedom of any burrs, no cold work hardening, and the parts are supplied absolutely free of oil and grease, ready to use for diffusion bonding. For cold work hardened sheets, residual stresses can be released, and the etch uniformity is poorer.

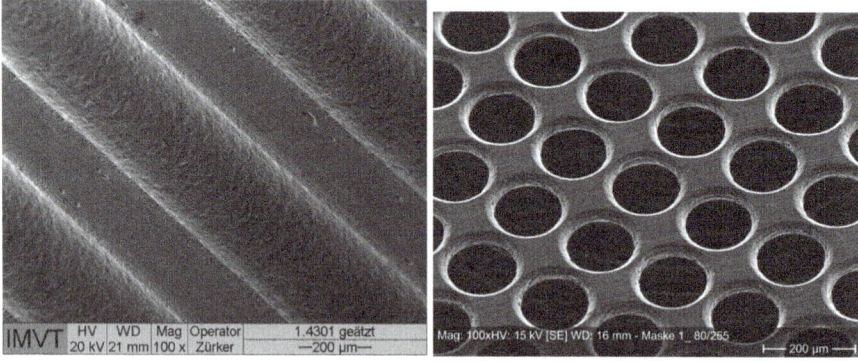

Fig. 9 Left: Etched microchannel, free of burrs, high surface roughness. Right: Microsieve, double-sided etched with etch rib

Waste materials containing heavy metal ions are costly to dispose. Therefore, it has to be considered whether microstructures are produced more efficiently, by etching or machining.

2.3 Laser Cutting

During laser cutting, the material is melted in a narrow, well defined area and expelled from the kerf by a focused jet of compressed gas.

The size of the heat-affected zone, burr formation and thermal distortion depend on a variety of parameters: Laser power, energy density, melting point of the material as well as its thermal conductivity, material thickness, beam cross-section and -quality and cutting speed as well as amount of surrounding material into which heat can dissipate.

For being able to cut very delicate microstructures, the beams cross section must be reduced to decrease the amount of heat deposited in the material, preventing thermal distortion (Fig. 10). Fiber lasers with the light conduction cable diameters down to 10 μm are available for that [22].

Due to the high energy density, all metallic materials, including refractory ones, can be cut. If necessary, inert gas must be used to prevent the reaction with atmospheric oxygen or nitrogen, e.g. for titanium or tantalum.

Fig. 10 Laser cut microbar, 0.5 mm in width, grooved wall profile. No significant burr formation

By means of laser cutting, arbitrarily shaped 2D structures can be created. Unlike machining, there is no strict relationship between the length of cut and the amount of material removed by laser cutting.

Only structures comprising the entire thickness of the material can be fabricated. Remaining webs require a suspension for fixation. Island-shaped structures cannot be realized.

What may appear to be a disadvantage can also be used to advantage: Since bottoms for passage separation are created by simple unstructured sheets, leakage due to material or structuring defects as for etching can be avoided. Moreover, the bottom thickness can be varied easily without any effort.

3 Joining Technologies

3.1 Diffusion Bonding

Diffusion bonding is the only joining technology that allows internal cross sections to be joined over the entire area without filler metal. Apparatus with excellent pressure resistance can be manufactured.

It is performed at about 80% of a materials melting temperature, calculated in Kelvin. Any work hardening is lost, but facilitates diffusion bonding by the formation of a new grain structure [23, 24]. The mechanical properties correspond to those of the soft annealed workpiece condition.

It has to be kept in mind that materials are not just the sum of their chemical components but also possess a mechanical and thermal history.

Dependent on the absolute temperature, a certain contact pressure is required to make intimate contact on atomic level between mating surfaces. A certain macroscopic deformation, which is difficult to predict and control, especially for microstructured devices, must be accepted [25, 26].

The joint is created by diffusion of atoms across the bonding planes. As a result, a monolithic part is formed by grain growth across where bonding planes cannot be distinguished anymore.

Since the diffusion coefficient depends exponentially on the bonding temperature and residual pores must be closed by volume diffusion, long dwell times are required [27].

Hence, there are many boundary conditions to be considered for successful diffusion bonding, depending on material: Surface roughness, presence of passivation layers, type of lattice, polymorphology of some sorts of materials, grain growth and unfavorable thermal effects such as the formation of precipitates at grain boundaries, which promote intercrystalline corrosion.

For micro process devices consisting of a large number of layers, and for increasing lateral dimension, additional issues from thickness tolerances from sheet rolling process may occur. The deformation behavior differs from parts consisting of a few layers only due to levelling of multiple surface roughnesses. The specification of a certain percentaged deformation to guarantee vacuum tightness is not sufficient. Instead, the deformation process should be controlled in-situ during diffusion bonding to avoid excessive deformation.

3.2   Laser Welding

Laser beam sources have undergone a rapid development in recent years. The cost per kilowatt has decreased enormously, which has promoted a considerable expansion and opened up new fields of application. Multi-kilowatt systems have now become widely available at reasonable cost.

Hence, laser welding is an economical joining technology with high processing speed. Welding speeds of several meters per minute are possible.

Due to the high energy density, nearly all metallic materials can be joint.

Additional material is not essential. However, high demands must be made on the accuracy of fit and freedom of gaps since air is an excellent thermal isolator.

The cross-section and shape of the weld seam can be varied within a wide range and adapted to the particular application (Fig. 11). At high welding speeds, the impact of heat conductivity of the material diminishes.

Due to melting and thermal shrinkage of material during solidification, seam bulging always occurs. For welding multilayer stacks at the perimeter, gaps may be formed. Hence, the welding technology in terms of sequence and direction is essential.

4   Interactions of Manufacturing and Joining Technologies

4.1   Diffusion Bonding and Material Properties

A sound diffusion bond is characterized by grain growth across bonding planes, so original sheets cannot be distinguished. Insufficient contact pressure, bonding temperature or dwell time as well as the existence of thermally stable passivation layers may prevent diffusion of atoms and closure of pores.

The impeding effect of passive layers must be emphasized. Their composition and stability depend on the type and quantity of alloying elements, while their thickness is predominantly determined by the thermal and environmental history. It can vary from below 10 nm to several hundreds of nanometer without visual change, causing varying results in diffusion bonding [28–32].

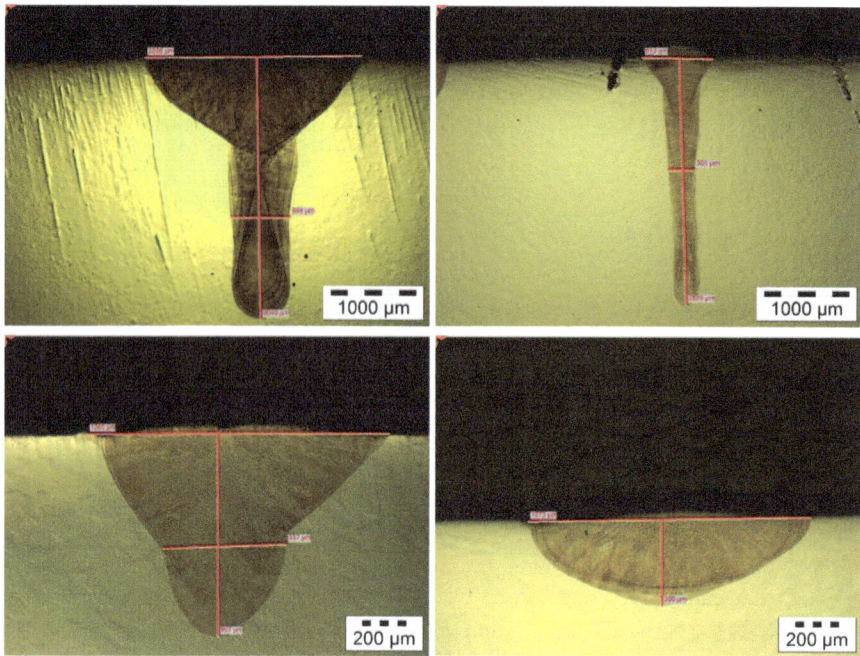

Fig. 11 Different cross sections of laser weld seams. Top left: High power, low speed. Tapered lead-in area due to heat conductivity. Top right: High power, high speed. Nearly rectangular cross section. Impact of heat conductivity diminished. Bottom left: Low to medium power, low speed. Triangular cross section. Bottom right: Low power, low speed. Flat seam with low welding depth. E. g. for joining porous inserts into thin sheet frames

Time dependent grain growth may degrade mechanical properties. Metallurgical impurities concentrate at grain boundaries and may form undesired precipitations, causing intercrystalline corrosion, excessive brittleness and decreasing impact strength [33, 34].

Polymorphic material possessing different crystallographic lattice exhibit varying diffusion coefficients with temperature by several orders of magnitude. The transformation of lattice leads to the formation of a new grain structure, promoting diffusion bonding extremely well.

4.2   Etching, Diffusion Bonding and Material Properties

Since the homogeneity of engineering materials is not comparable to that of silica wafers, the question arises about material defects (pores, nonmetallic inclusions), minimum possible residual bottom thicknesses and etching breakthroughs, which may cause leakage between different passages, especially in micro process technology.

The origin of the material and melting technologies but also comparable material specifications originating from different historical epochs have to be considered. For example, austenitic stainless steel 316L (1.4404) differs from 316Ti (1.4571) only in a maximum carbon content of 0.03% instead of 0.08%. That seems to be a small difference, however neglecting that the density of carbon is about one fourth of iron, underestimating its content in atomic percent. In 316Ti, 0.7% titanium is added to absorb carbon as titanium carbides due to its higher affinity of titanium. Otherwise, the formation of chromium carbides would reduce corrosion properties. The background is that melting technologies at that time were not able to further reduce the carbon content at reasonable costs. 316L does not contain titanium. Nevertheless, the content of nonmetallic inclusions is lower, decreasing the risk of etch break-troughs.

A proven way to prevent leakage is to stack etched sheets back-to-back. This eliminates the possibility of pores to superimpose each other.

However, it leads to varying cross sections for different layer, leading to varying contact pressures during diffusion bonding. Therefore, appropriate bonding during diffusion bonding may be prevented for selected layers.

4.3 Laser Cutting and Diffusion Bonding

To achieve a high surface-to-volume ratio of micro process devices, only narrow webs are desired (Fig. 12). If only a small volume of material is left where heat can dissipate, thermal distortion is an issue. In order to be able to successfully laser cut delicate structures, the focus diameter of the laser beam must be reduced as much as possible. Since the area decreases quadratically with decreasing diameter, considerably lower laser powers are required for the same or higher energy density. On the other hand, the beam cross-section must be sufficient to melt volumes corresponding to the material thickness and for being able to expel them from the kerf by compressed gas.

The stability of webs must be sufficient to be suspended by bars at the end to fix them in position. After stacking and diffusion bonding, the ends of the device must

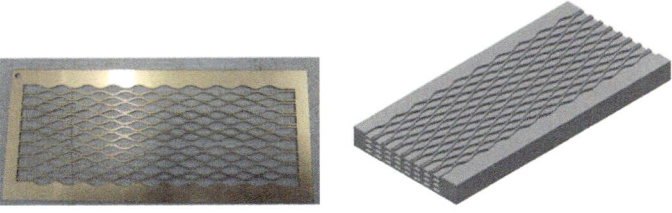

Fig. 12 Left: Laser cut structure (115 × 50 mm^2) with suspended webs (1 mm in width). Right: Stacked, diffusion bonded and eroded

be eroded to expose the microchannels. Alternatively, microchannel structures under a certain angle can be cut and stacked alternating, forming grid-like structures.

By using sheet material of different thickness, the bottom thickness between different passages can be tailored very easily without any effort and risk of leakage.

4.4 Laser Welding for Micro Process Devices

Laser welding can be used for integration of microstructured components into housings as well as for joining of multilayer stacks due to the flexibility of the weld seam cross section. Microchannel structures can significantly alter the dissipation of heat. High welding speeds should be preferred (Fig. 13) [35].

Again, due to seam bulging and shrinkage, gaps between different layers may arise when laser welding multilayered stacks. Perfect clamping, sequence and direction of welding are essential.

For through welding of multilayered assemblies, appropriate clamping technology is required for guaranteeing process reliability (Fig. 14).

At different welding levels, masking trenches must be shifted for not crossing each other for subsequent layers, which requires lots of design work [36].

In the case of several fluid passages, there is the dilemma of either performing a leakage test with appropriate adapters after each newly welded layer or testing the completely finished device. In this case, however, repair of buried welds is not possible. Therefore, reliable process control is very important.

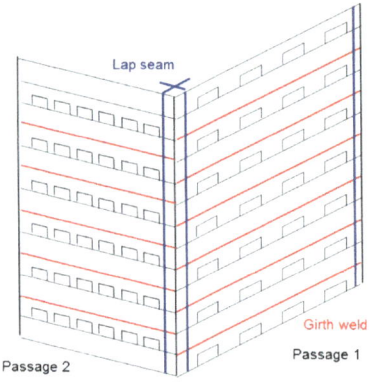 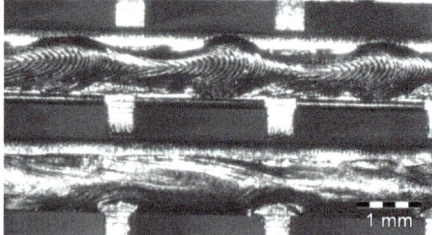

Fig. 13 Left: Joining of multilayered stack by laser welding and separation of different passages. Right: Detail of microstructured stack. Top weld seam: v = 2.4 m/min, periodically varying width due to microchannels. Below: 6 m/min welding speed

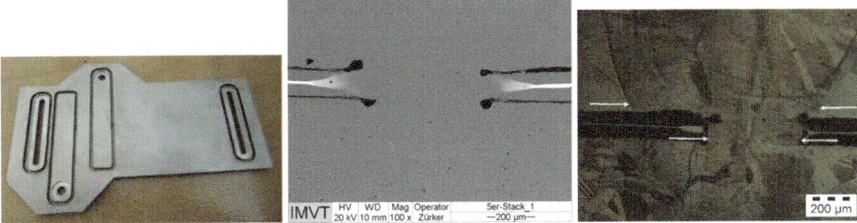

Fig. 14 Left: Through welding of multilayered assemblies. Weld seam bulging must be masked by etched or machined trenches. Middle: Integration of a 12 μm palladium membrane between 50 μm etched sieves and sheets 1 mm in thickness. Right: Weld necking due to thermal isolation between multiple layers due to gap by insufficient clamping

5 Conclusions

Micro process devices can be manufactured using various structuring and joining technologies. However, there are many details to be considered that are usually neglected in the macroscopic world.

Very tight specifications must be met for both, diffusion bonding and laser cutting as well as laser welding.

Often, there are factors that have not been considered in advance or that depend on each other, which may cause problems or lead to failure of micro process devices.

Skilled stuff, familiar with micro process technology is a prerequisite for success.

These are reasons why the transfer of technology to commercial companies requires close supervision.

Acknowledgements The financial support from the Helmholtz Program MTET (Materials and Technologies for the Energy Transition) is gratefully acknowledged.

References

1. Gietzelt, T., Hüll, A., Toth, V., Messerschmidt, F., Thelen, R.: Impact of scratch depth on vacuum tightness of diffusion bonded parts. Mat.-wiss. u. Werkstofftech. **49**, 185–192 (2018). https://doi.org/10.1002/mawe.201700154
2. nashira Hard Metals: Eigenschaften des Hardmetalls. Technische Merkmale. https://www.nashira-hm.it/de/die-eigenschaften-des-hartmetalls/. Last accessed 24 May 2022
3. Sailer, T.: Ultrafeinkörnige Hartmetalle mit Co-Binder und alternativen Bindersystemen - Korrelation von Mikrostruktur und mechanischem Verhalten unter monoton ansteigender und zyklisch wechselnder Beanspruchung. Ph.D. thesis, Univ. Erlangen-Nürnberg (2002)
4. Ceratizit Group: Carbide is a matter of confidence. https://www.ceratizit.com/int/de/media/downloads.html#/sections/item0-Downloads_Hard_Material_Solutions. Last accessed 24 May 2022
5. Schatt, W. Wieters, K.-P., Kieback, B.: Pulvermetallurgie, 2. Ed., Springer, Berlin, Heidelberg, New York (2007), chap. 17.2, 510ff. ISBN-13 978-3-540-23652-8

6. Moldino, Quickfinder-Vollversion 1.16.0.3. https://www.moldino.eu/. Last accessed 24 May 2022
7. Bandapalli, C., Sutaria, B.M., Bhatt, D.V.P., Singh, K.K.: Tool wear analysis of micro end mills—uncoated and PVD coated TiAlN & AlTiN in high speed micro milling of titanium alloy—Ti-0.3Mo-0.8Ni. Proc. CIRP. **77**, 626–629 (2018). https://doi.org/10.1016/j.procir.2018.08.191
8. Hollek, H., Schier, V.: Multilayer PVD coatings for wear protection. Surf. Coat. Technol. **76–77**, 328–336 (1995). https://doi.org/10.1016/0257-8972(95)02555-3
9. Bienk, E. J., Reitz, H., Mikkelsen, N. J.: Wear and friction properties of hard PVD coatings. Surf. Coat. Technol. **76–77**, 475–480 (1995). https://www.sciencedirect.com/science/article/pii/0257897295024980
10. Chowdhury, S., Beake, B.D., Yamamoto, K., Bose, B., Aguirre, M., Fox-Rabinovivh, G.S., Veldhuis, S.C.: Improvement of wear performance of nano-multilayer PVD coatings under dry hard end milling conditions based on their architectural development. Coatings **8**, 59 (2018). https://doi.org/10.3390/coatings8020059
11. GDE-Werkzeuge GmbH. https://www.gde-werkzeuge.de/. Last accessed 24 May 2022
12. HAM Präzision. https://ham-tools.com/. Last accessed 24 May 2022
13. Mössner. https://moessner-gmbh.com/de/zerspanen. Last accessed 24 May 2022
14. Medidia. https://www.medidia.eu/. Last accessed 24 May 2022
15. Wan, Z., Li, Y., Tang, H., Deng, W., Tang, Y.: Characteristics and mechanism of top burr formation in slotting microchannels using arrayed thin slotting cutters. Precis. Eng. **38**, 28–35 (2014). https://doi.org/10.1016/j.precisioneng.2013.06.008
16. Herz Ätztechnik. https://www.aetztechnik-herz.de/. Last accessed 24 May 2022
17. Fotofab. https://fotofab.com. Last accessed 24 May 2022
18. Micrometal. https://www.micrometal.de/. Last accessed 24 May 2022
19. Precision Micro. http://www.precisionmicro.de. Last accessed 24 May 2022
20. etchform. https://www.etchform.nl. Last accessed 24 May 2022
21. Gandhi, S.V., Chanmanwar, R.M.: A study of variation in MRR influenced by work piece positioning on copper and stainless steel during wet chemical machining. In: ATSMDE2017 (2017). https://papers.ssrn.com/sol3/papers.cfm?abstract_id=3101587
22. Trumpf GmbH + Co KG, Trufiber Compact Series. https://www.trumpf.com/de_DE/produkte/laser/faserlaser/trufiber-p-compact/. Last accessed 24 May 2022
23. Harumoto, T., Yamashita, Y., Ohashi, O., Ishiguro, T.: Influence of cold rolling on diffusion bondability of SUS316L stainless steel sheets. Mater. Trans. **55**(3), 633–636 (2014). https://doi.org/10.2320/matertrans.M2013405
24. Katoh, M., Sato, N., Shiratori, T., Suzuki, Y.: Reduction of diffusion bonding temperature with recrystallization at austenitic stainless steel. ISIJ Int. **57**(5), 883–887 (2017). https://doi.org/10.2355/isijinternational.ISIJINT-2016-693
25. Saranam, V.R., Paul, B.K.: Feasibility of using diffusion bonding for producing hybrid printed circuit heat exchangers for nuclear energy applications. Procedia Manuf. **26**, 560–569 (2018). https://doi.org/10.1016/j.promfg.2018.07.066
26. Southall, D., Le Pierres, R., Dewson, S.J.: Design considerations for compact heat exchangers. United States: American Nuclear Society - ANS (2008). https://inis.iaea.org/search/search.aspx?orig_q=RN:42096311
27. Xue, K., Tian, W., Yan, S., Li, H., Li. P.: Variations in mechanical properties of RAFM steel under vacuum diffusion welding with pre-deformation and subsequent heat treatment. Fusion Eng. Des. **152** (2020). https://doi.org/10.1016/j.fusengdes.2020.111470
28. Takahashi, Y., Nakamura, T., Nishiguchi, K.: Dissolution process of surface oxide film during diffusion bonding of metals. J. Mater. Sci. **27**, 485–498 (1992). https://link.springer.com/article/https://doi.org/10.1007/BF00543942
29. Machet, A., et al.: XPS and STM study of the growth and structure of passive films in high temperature water on a nickel-base alloy. Electrochim. Acta. **49**, 3957–3964 (2004). https://doi.org/10.1016/j.electacta.2004.04.032

30. Shih, C.-C., et al.: Effect of surface oxide properties on corrosion resistance of 316L stainless steel for biomedical applications. Corr. Sci. **46**, 427–441 (2004). https://doi.org/10.1016/S0010-938X(03)00148-3
31. Hakiki, N.E., Montemor, M.F., Ferreira, M.G.S., da Cunha Belo, M.: Semiconducting properties of thermally grown oxide films on AISI 304 stainless steel. Corr. Sci. **42**, 687–702 (2000). https://doi.org/10.1016/S0010-938X(99)00082-7
32. Gietzelt, T., Toth, V., Weingärtner, T.: Impacts of layout, surface condition and alloying elements on diffusion welding of micro process devices. Materialwiss. Werkstofftech. **9**, 1070–1084 (2019). https://doi.org/10.1002/mawe.201800197
33. Zhu, Z.-Y. et al.: Effect of aging treatment on intergranular corrosion properties of ultra-low iron 625 alloy. Int. J. Corros. (2019). Article ID 9506401. https://doi.org/10.1155/2019/9506401
34. Bansod, A.V., Patil, A.P., Moon, A.P., Khograga, N.N.: Intergranular corrosion behavior of low-nickel and 304 austenitic stainless steels. J. Mater. Eng. Perform. **25**(9), 3615–3626 (2016). https://doi.org/10.1007/s11665-016-2221-2
35. Gietzelt, T., Eichhorn, L., Wunsch, T., Schorle, C., Kaut, M., Dittmeyer, R.: Laser welding of multilayer stacks made of thin-sheet material for the manufacture of microstructured devices for process engineering. Chem. Ing. Tech. **85**(10), 1624–1631 (2013). https://doi.org/10.1002/cite.201200181
36. Boeltken, T., Wunsch, A., Gietzelt, T., Pfeifer, P., Dittmeyer, R.: Ultra-compact microstructured methane steam reformer with integrated Palladium membrane for on-site production of pure hydrogen: experimental demonstration. Int. J. Hydrogen Energy **39**(31), 18058–18068 (2014). https://doi.org/10.1016/j.ijhydene.2014.06.091

Applications

Design Management and Advanced Technologies in the Development and Production of a Modern Recreational Craft

Vitor Carneiro⬤, Francisco Cálão, Jorge Lino Alves⬤, and Augusto Barata da Rocha⬤

Abstract The Recreational Craft Production (RCP) sector has undergone important advances in the last decade, mainly due to the incorporation of design management and advanced technologies in product development, prototyping and production. The RCP sector, that represents in Europe more than €20 billion in annual revenue, is a highly competitive sector with important challenges related to design, productivity, competitiveness, safety and environmental requirements. This paper describes the complete stages of product development/production of a modern craft using design management concepts, CAD/CAM/CAE technologies, FEM simulations for structural stability and flow simulation of the friction interface hull/sea water. A modern multi-step hull is developed to reduce friction and the corresponding pressure distribution is analysed and compared to the corresponding traditional hull (without step). The coordinates of the centre of mass, centre of buoyancy and waterline are calculated. The longitudinal and transversal stability is analysed and the Gz curve presented, for both cases. The calculations show that the introduction of a stepped hull has a major influence on the pressure distribution under the hull and on the final coordinates of the centre of buoyancy. The mould design of both hull and deck is

V. Carneiro (✉)
Faculty of Engineering, University of Porto, Rua Dr. Roberto Frias s/n, 4200-465 Porto, Portugal
e-mail: up200706976@fe.up.pt

F. Cálão
Nautav Lda, Zona Industrial da Mota, Rua 12, Lote nº66, 3830-527, Gafanha da Encarnação, Portugal

Zona Industrial da Mota, Rua 12, Lote nº66, 3830-527 Gafanha da Encarnação, Portugal

J. L. Alves · A. B. da Rocha
Department of Mechanical Engineering, Faculty of Engineering, University of Porto, Rua Dr. Roberto Frias s/n, 4200-465 Porto, Portugal
e-mail: falves@fe.up.pt

A. B. da Rocha
e-mail: abrocha@fe.up.pt

J. L. Alves
INEGI, Campus da FEUP, Rua Dr. Roberto Frias nº400, 4200-465 Porto, Portugal

107

presented and a technique for interference and clearance evaluation for demoulding, before CNC machining of the moulds is applied. Additive manufacturing, used to produce scale models of the final craft, demonstrates major benefits to improve design and to discuss commercial and technical issues. Different alternatives of manufacturing processes (hull and deck) are analysed during the production phase. Finally, the application of design management, additive manufacturing and advanced technologies allows a reduction of time-to-market, the successful achievement of the final reviewed specifications of the craft and the increase in competitiveness and productivity of the final product.

Keywords Design management · Product development · Recreational craft · SME's

1 Introduction

One of the first records of the use of recreational crafts dates from the beginning of the seventeenth century in Holland, when crafts were created for sports and leisure purposes, for use by nobles and kings with the intention of celebrating the arrival of large merchandising ships (Kobbé 1914).

These recreational crafts were initially built in wood, with classical techniques of naval architecture. The first patent related to fiberglass was awarded to the Prussian inventor Hermann Hammesfahr (1845–1914) in the United States of America in 1880 (Hammesfahr 1880; Mitchell 1999). A suitable resin for combining the fiberglass with a plastic to produce a composite material was developed in 1936 by DuPont. The first produced composite craft is credited to Ray Greene of Owens Corning in 1937 (Marsh 2006).

The sales of recreational crafts began to have some commercial success only from the mid-nineteenth century, with the beginning of nautical races in Europe and the United States of America (Boats 2018). However, the Second World War stagnated de development and production of the fiberglass recreational crafts, despite their crucial role, when hundreds of crafts crossed the Channel to the French port of Dunkirk between 27 May and 4 June 1940. Fishing boats, pleasure yachts and lifeboats—known as the little ships—were all pressed into service to rescue hundreds of thousands of British, French and Belgian soldiers who had been forced back to the coast in the face of the German advance across Europe. This event was officially called "Operation Dynamo" and proved to be a pivotal moment of the Second World War (Press Association 2015).

The beginning of high competition nautical sports helped to popularize this composite, due to the advantages it presents, making the craft lighter and cheaper and consequently more competitive (Mitchell 1999).

Historically, Portugal has always been recognized worldwide as a country with very strong connections to nautical activities, from the time of the Discoveries (1415–1543), a period that allowed a revolution in the maritime industry, to the present day.

In 1964, Manuel Alves Barbosa, was invited by the King of Morocco to participate in a powerboat race. Fascinated by the fiberglass technology and potential of this new method of shipbuilding, highly resistant, maintenance-free and with an unparalleled versatility of innovation and series construction, Manuel Alves Barbosa started the negotiations that culminated in the creation of the Barbosa e Sciacca Lda. shipyard in Aveiro. This was the first Portuguese shipyard to produce fiberglass crafts and one of the first at European level, later called Riamar.

In the last decade the nautical market has seen a remarkable growth. As it is an extremely competitive market, in which new models are manufactured annually, equipped with innovative technologies and appealing design, companies in this sector are forced to periodically introduce new models on the market. It is in this context of growth of this industry that Nautav (owner of the Riamar), in partnership with the DesignStudio of the Faculty of Engineering of the University of Porto (FEUP), started a new approach of craft design and production combined with design management procedures, with the development of a new models, in order to expand its currently offer currently.

This paper is structured as follows. In the first section an integrated approach of design management is presented with the identification of the main topics and parameters for design management during the product development process in the context of Small and Medium-Sized Enterprises (SMEs). In the second section, the stages of the product development of a modern recreational craft are presented. A modern multi-step hull is developed to reduce friction, to increase performance and reduce consumption. In the third section stability calculations and FEM flow simulations of the pressure interface hull/sea water were performed and analysed. A comparison between the developed multi-step hull and a traditional hull (without step) is analysed. In the fourth section, the prototypes production is described. In the fifth section the mould fabrication and corresponding craft production is described. Finally the main conclusions are presented.

2 Design Management Integration

Design management in the context of SMEs is particularly important and challenging. These organizations face numerous financial and human resource constraints; and sometimes they lack the design skills and experience they need (Oakley 1982) which means they must find those competences and resources externally. In addition, and unlike large companies, many SMEs have no formal methodology to frame design development (Berends et al. 2011) and run on outdated management models (Alarcon et al. 2015) designed to be rigid and isolated which block their desired strategy from becoming reality.

Given the value that design can contribute to business performance for SMEs (Berends et al. 2011; Bruce et al. 1999; Chiva and Alegre 2009), optimizing the product development process and its management have become strategic imperatives. The design process, however, requires input from multiple disciplines, and may consist of several integrated methodological approaches. SMEs may not have the required expertise in house, and so the company and external project contributors may conduct the process jointly.

In a previous publication (Carneiro et al. 2021) the authors presented a survey and analysis of the literature identifying the key parameters, dependencies, and connections involved in SME design management (at the operational level) during the product development process. It was shown that parameters of design management involved in the SME product development process are complex, as are the connections between them. At the heart of the SME, product development process are a complex sets of design management parameters and dependencies that require careful consideration (Fig. 1). Each of the five aspects of design management comes with a set of relevant subtopics directly or indirectly related to each other. Their connections become even more intricate once the sub-connections that also occur between the subtopics of different aspects are considered—all of which are combined to create a complex web of interdependencies. For example, in the "Managing Design Projects" group, the "Project Design" subtopic is sub-connected to the "Design Coordination" and "Design Skills" subtopics of the "Design Management Capabilities" group (marked with a dotted yellow line). To exemplify this complexity, and render it visible, we have chosen to depict potential connections and sub-connections among subtopics using a single, continuous, graded line between groups. More sub-connections exist, attesting the complexity inherent in the relationships among all these aspects of design management.

The process of innovation in the manufacturing of new crafts at Nautav company was accompanied by a detailed analysis and internal discussion of each of the parameters, described in Fig. 1, to assess their potential and constraints. In a pilot study, each parameter was evaluated quantitatively and qualitatively, allowing a realistic

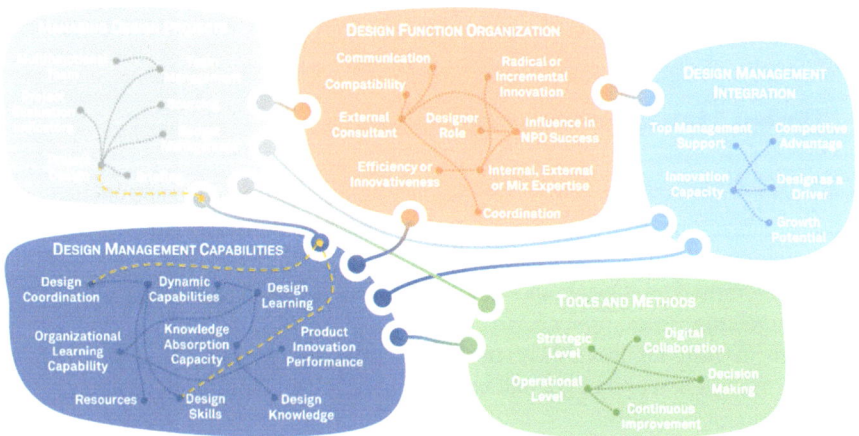

Fig. 1 Map of the main topics and parameters for design management during the product development process in the context of SMEs

image of the company's strengths and weaknesses in each of the multiple areas that constitute the design management. This methodology allowed the identification of critical sectors of this innovation process and taking measures to mitigate them. It was identified that the Design Function Organization was the most critical area that had to be reinforced with higher commitment at the top management level.

This pilot study also made possible to obtain a clearer and more structured image of the main topics and parameters for design management during the product development process in the context of SMEs, which will allow the development of a future structured methodology for the implementation of a design management policy in SMEs.

3   Recreational Craft Product Development

Recreational crafts are classified according to category, craft typology and propulsion system (DGRM 2018). The craft developed in this project is category C, considered suitable for winds of force equal to or less than 6 on the Beaufort scale and waves with an indicative height equal to or less than 2 m; the typology of the craft is constituted by deck and hull joined at the boat fender zone (Fig. 2), with structural pavement and equipped with an outboard engine.

The hull is the main structure of most marine crafts, allowing the boat to float and withstand the forces resulting from the weight of the rest of the structure and crew. The deck is the permanent cover over the hull, which, in addition to protecting against the outside elements, also provides structural reinforcement to the hull. On recreational boats, the deck is the social area where passengers spend most of their time. The hull and deck are joined in the area of the fender, a rubber, wood or stainless steel strip that has the function of absorbing shocks between crafts or between the craft and the pier(s).

Considering the target market, Nautav defined the specifications with all the project requirements, of which the following stand out:

Fig. 2 Hull, deck and fender zone of a recreational craft

- Length of the craft (not including the engine) from bow to stern of 6.5 m and a maximum width of 2.5 m;
- Possibility of assembling several deck versions with the same hull, allowing the creation of a new line and expanding the offer;
- Craft equipped with an outboard engine from 130 to 250 hp;
- Deck coated with non-slip synthetic teak;
- Finishing of the boat with gelcoat to avoid painting;
- Production of the craft through manual moulding or, alternatively, by infusion, without ever losing the quality of construction and finishing that characterize Riamar crafts. Through this fully automated process, it is possible to minimize the joints between hull and deck and the finishing work. In addition, this process is fully automated, minimizing human error;
- Finally, a hull with a double step. The step (Fig. 3) is a transversal cavity that runs along the hull from side to side, and comes high enough on the side of the craft to reach above the waterline. As the craft moves forward, the side air inlets of the cavity allow to create a "cushion" of air bubbles into the cavity creating a low pressure zone and as result drag forces that the water causes on the hull decrease (Rudow 2012). The advantages of using this technology consist of a reduction in drag and friction forces, resulting in a greater ability of the craft to reach higher speeds compared to a craft without a step (traditional hull) and to obtain better efficiency in fuel consumption.

The modelling of the craft was performed using 3D CAD software, through surface and solid modelling. Initially, each component of the craft was individually modelled, followed by assembly and respective analysis of minimum and maximum clearances and interference between the various components (Fig. 4). Additionally, the moulds and inserts of the hull and deck were also modelled and the respective assembly was carried out. This analysis allowed simulating the demoulding process, verifying the existence of interferences and their causes, thus avoiding and anticipating errors in the production phase (Fig. 5). This phase is complex with multiple iterations of modelling, assembling, simulation and analysis, until the desired result is obtained.

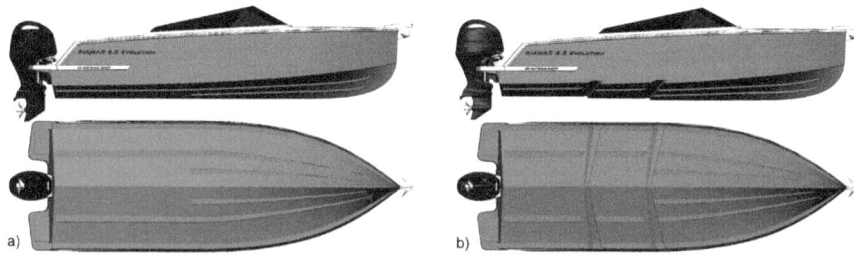

Fig. 3 Traditional hull (**a**) and multi-stepped hull (**b**)

Fig. 4 3D modelling process: **a** individual modelling of components; **b** components assembly; **c** analysis of interferences and clearances of the various components

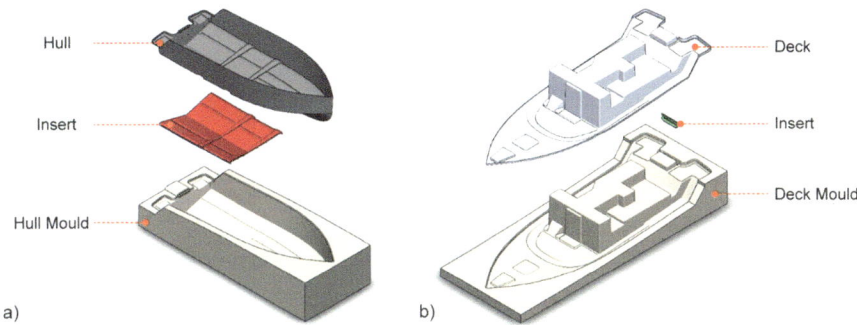

Fig. 5 3D model of the mould and insert of the hull (**a**) and deck (**b**) to simulate the demoulding process

4 Stability Calculations and Flow Simulation (FEM)

The static analysis of longitudinal and transversal stability makes it possible to understand the behaviour of the craft in the water, namely its centre of gravity (G), the waterline, centre of buoyancy (B) and the maximum heel supported by the craft. The dynamic analysis of the craft, using the finite element method (FEM), allowed the simulation of the boat wake, pressure distribution on the surface of the hull and the drag force caused by the water when in contact with the hull. To understand the analyses carried out, it is necessary to understand the different concepts related to the longitudinal and transverse stability of a craft.

4.1 Basic Principles

4.1.1 Archimedes' Principle, Displacement, Centre of Gravity and Centre of Buoyancy

Archimedes' principle states that the upward buoyancy force that is exerted on a body immersed in a fluid, whether fully or partially, is equal to the weight of the fluid that the body displaces. Archimedes' principle is a law of physics fundamental to fluid mechanics (Oliveira and Lopes 2013).

The weight of the fluid displaced by a craft is known as *Displacement*, and the displaced water creates an upward force, or buoyancy force, which is equal to the weight of the craft. The displaced water has a centre of mass, or *Centre of Buoyancy* (B), which varies according to the shape of a craft's hull and keel. The weight of a craft is distributed along its length, pushing the entire craft downwards. All the weight acts downwards through its *Centre of Gravity* (G).

4.1.2 Longitudinal Stability

In static equilibrium, the centre of gravity (G) and the centre of buoyancy (B) are aligned with z-axis (Fig. 6).

4.1.3 Transversal Stability, Angle of Heel, GZ Curve, Angle of Vanishing Stability (AVS)

To keep a craft stable in the water and prevent it from leaning over, the centre of gravity needs to be as low as possible. When the craft is in the upright position, the centre of gravity (G) is aligned vertically with the centre of buoyancy (B) and there's no righting lever (Gz). If the craft heels to the wind or waves, the centre of buoyancy (B) will move to laterally and a righting lever (Gz) is generated (Fig. 7).

Figure 8 shows a typical Gz curve (or Righting Moment Curve or Curve of Static Stability). When the craft heels, the righting lever (Gz) will increase to a maximum (60° of heel in Fig. 8). If the craft continues to heel the righting lever (Gz) starts to reduce until it reaches zero again (130° of heel in Fig. 8). This point is called the

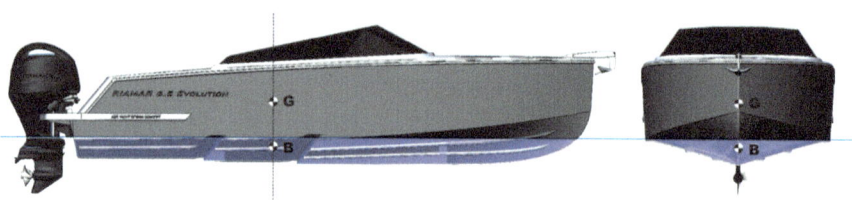

Fig. 6 Centre of gravity (G) and the centre of buoyancy (B) aligned with z-axis

Fig. 7 Centre of gravity (G), Centre of Buoyancy (B) and Righting lever (Gz). Adapted from Simpson (2022)

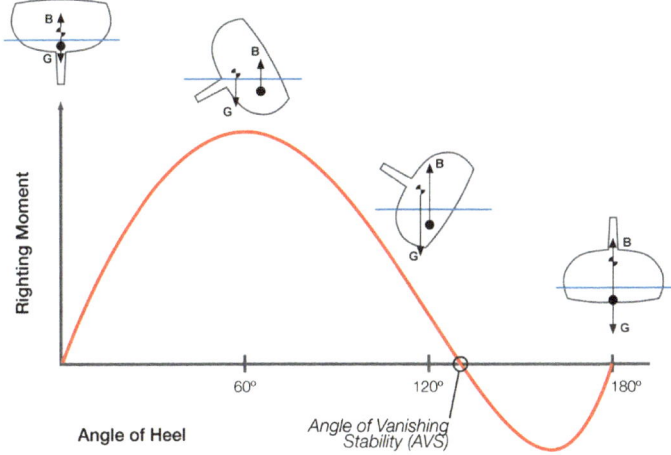

Fig. 8 Gz curve and he angle of vanishing stability (AVS). Adapted from Simpson (2022)

Angle of Vanishing Stability (AVS), also known as the Limit of Positive Stability (LPS). If the angle of heel exceeds the AVS the righting lever (Gz) will become negative and will act as a capsizing lever rather than righting lever.

Righting Moment Curves (Gz curves) are published by craft manufacturers to show the stability characteristics of their crafts designs. In Europe, the Recreational Craft Directive (RCD) states that recreational crafts between 2.5 and 24 m must carry builder's plates to categorize their boats in either Category A (Ocean), B (Offshore) or C (Inshore) and meet minimum standards of stability.

4.2 Longitudinal Stability Calculations

Using CAD software, the centre of gravity (G) was calculated using fiberglass density of 2440 kg/m^3 and the weight of a 200 hp engine was also considered. The total estimated weight of the boat was 1531 kg. Figure 9 shows the centre of gravity (G) along x and z axis and their coordinates (mm).

The craft's waterline was calculated by equalizing the buoyancy force and the craft's' total weight. The coordinates of the centre of buoyancy (B) are obtained from the centre of gravity of the displaced water (Fig. 10).

For longitudinal stability on a craft it is necessary that the centre of gravity and the centre of buoyancy are vertically aligned, while the keel is parallel to a horizontal plane. Comparing the previous two figures, it can be seen that the x-axis position of the centre of gravity must be pushed forward ~147 mm so that the craft is in a horizontal position and longitudinal stability is guaranteed. With the addition of other essential components of the craft such as the mooring system, cabin furniture and navigation systems it is possible to correct this difference.

The following graph (Fig. 11) shows the evolution of the buoyancy force and corresponding centre of buoyancy (B) with increasing water line levels. The blue horizontal

Fig. 9 Centre of gravity (G) of the designed craft (dimensions in mm)

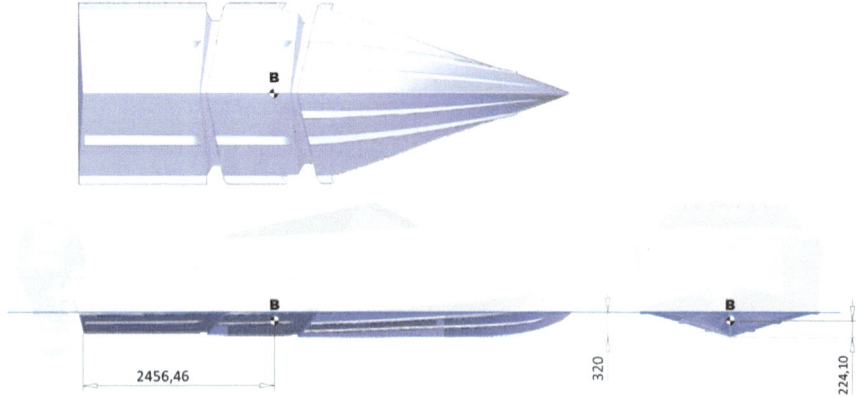

Fig. 10 Calculation of the waterline and the centre of buoyancy (B) (dimensions in mm)

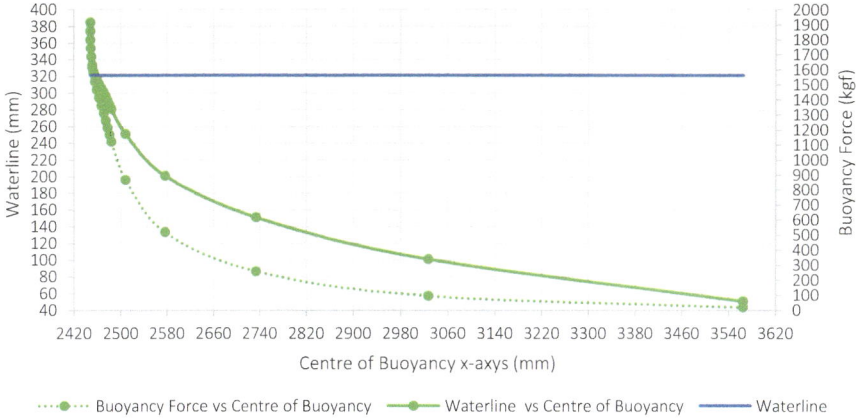

Fig. 11 Longitudinal stability of the designed craft

line corresponds to the craft's final waterline (320 mm of draft) and corresponding centre of buoyancy (B).

4.3   Transversal Stability Calculations

The calculation of the transversal stability allowed to obtain the Angle of Vanishing Stability (AVS). The Righting Lever (Gz) and the corresponding Righting Moment (RM) values were calculated for each angle of heel. The craft was rotated around its centre of gravity (G), through an axis parallel to the x-axis (longitudinal direction of the craft). For each heel angle the waterline was calculated by equalizing the buoyancy force with the weight of the craft. The following graph (Gz Curve) (Fig. 12) shows the Righting Moment (RM) for each angle of heel. As it can be seen in the graph, the AVS is 89°.

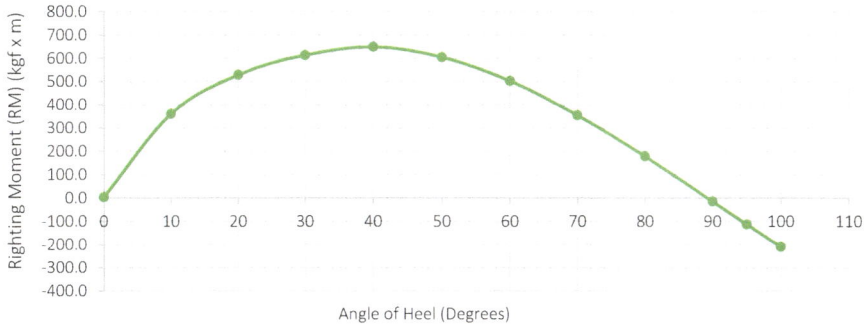

Fig. 12 Transversal stability (Gz curve) of the multi-step hull

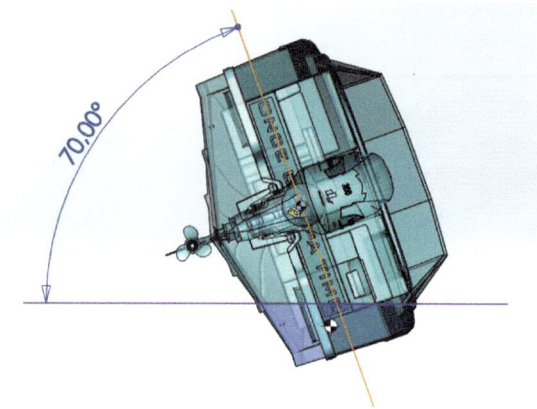

Fig. 13 Maximum angle of heel of the designed craft

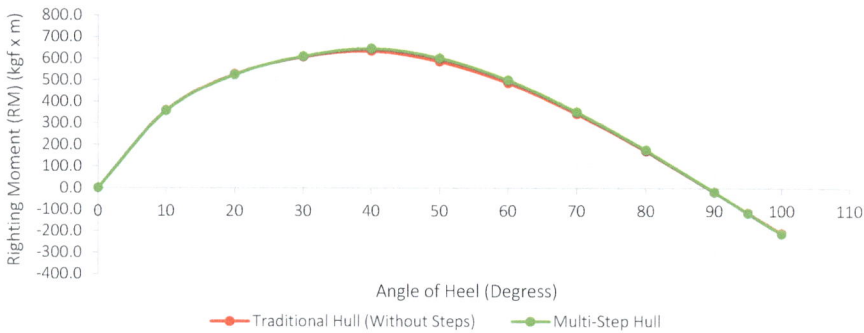

Fig. 14 Comparison of transversal stability (Gz curve) of the multi-step hull and the corresponding traditional hull

However, due to the craft pit edge, the maximum angle of heel is lower, it is limited to 70°, as shown in Fig. 13. If the crafts exceeds the heel angle of 70°, water will start overflowing the edge.

Figure 14 shows a comparison of the righting moment of the multi-step hull with the corresponding traditional hull (Gz Curve). As it can be seen, from a transversal stability point of view, the two hulls are very similar.

4.4 Dynamic Analysis

This chapter describes the dynamic analysis of the craft's performance in open water. For comparative and analysis purposes, both the hull with the double step and without the step, known as the traditional hull, were simulated. For the analysis, the add in

Flow Simulation, from the SolidWorks software, was used to obtain the CFD simulations, introducing parameters related to atmospheric conditions and the behaviour of the fluid.

4.4.1 Simulation Parameters

To carry out the simulations the following input parameters were defined (Table 1).

The creation of an automatic mesh with advanced refinement was also selected (thinner in the impact zones between the hull and the water) and the simulation volume comprises a length of 33 m, a width of 22 m and a height of 5 m. In order to simulate the trim effect caused by the engine propelling force, an inclination of 5° was assigned (Ghadimi et al. 2014). The total simulation time was respectively 1 h and 13 min for the traditional hull and 1 h and 15 min for the multi-step hull.

4.4.2 Results

Figure 15 presents the flow disturbance caused by the craft displacement at 16 knots. As it can be seen, the traditional hull creates higher disturbance than the multi-step hull. This means a lower drag of the multi-step hull leading to increased performance and fuel consumption reduction. In this analysis wave effect of water surface was not considered.

In Fig. 16 it is possible to observe the pressure distribution in the hull surface resulting from the 16 knot speed. The traditional hull presents a smother pressure distribution in middle region of the hull. However, the multi-step hull presents lower pressures (green regions) at the peripheral region of the hull which creates a desired air suction effect and resulting air bubble cushion that reduces drag and increases

	Parameters	Type/value
Table 1 Parameters and values used in the simulations	Flow analysis	External
	Total analysis time	10 s
	Output time	1 s
	Type of flow	Free surface
	Fluids	Air and water
	Regime	Laminar and turbulent
	Thermal condition on the surface	Adiabatic walls
	Pressure	101,325 Pa (1 atm)
	Temperature	293.15 K (20 °C)
	Speed	16 Kn (8.23 m/s)
	Water line	0.290 m (traditional hull) 0.320 m (multi-step hull)

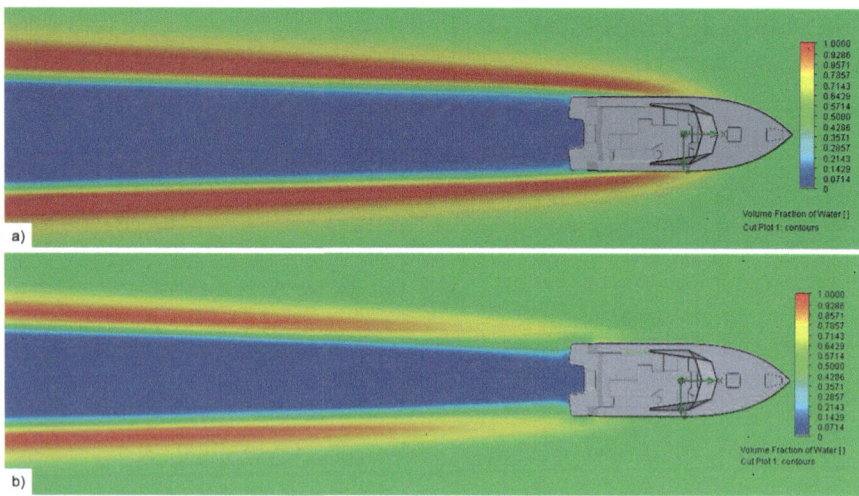

Fig. 15 Water flow disturbance at a speed of 16 knots: **a** traditional hull; **b** multi-step hull

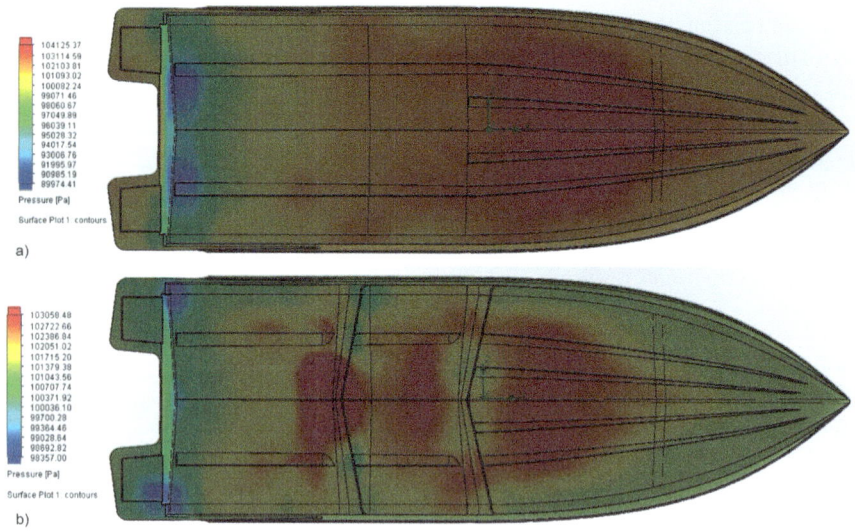

Fig. 16 Hull pressure distribution at a speed of 16 knot: **a** traditional hull; **b** multi-step hull

performance. It is also observed that the lowers pressures (blue regions) in the stern region are more peripheral in the multi-step hull and more central in the traditional hull leading to increased transversal stability for the multi-step hull.

5   Prototypes Manufacturing

To validate the design and global structure of the craft, prototypes were produced by 3D printing using the fused filament extrusion (FFF) manufacturing process, at a scale of 1:20. For the printing of the prototypes, a Prusa i3 MK3S printer was used with a 0.4 mm nozzle fed with PLA (*Polyactic Acid*) filament of 1.75 ± 0.05 mm in diameter, where a layer resolution of 0.15 mm was used. It is also important to highlight the fact that the craft is printed in two separate parts, due to its size being greater than the capacity of the printing table (Fig. 17a, b). In total, it took 267 h of printing and 1765 g of filament. FFF technology allows the production of parts with good quality and mechanical strength. In addition, and by the same process, moulds (positive and negative) of the hull and deck were also produced at a scale of 1:40 (Fig. 17c).

For the 3D printing of the engine, PolyJet technology was used, which allows the production of thin-walled parts, complex geometries and intricate details. Additionally, it enables to create parts with movements with a single print, which is the case of the engine (two degrees of freedom: steering and trim). The engine was printed on an EDEN 260 V equipment (Stratasys), with deposition of 16 μm light-curing acrylic resin layers. Vero Gray material was used, with a total printing time of 1.5 h, with a consumption of 76 g of construction material and 51 g of support material (Fig. 18).

The successive iterations of the hull and deck design supported by prototypes manufactured by 3D printing allowed to reach optimized final versions, both in terms of design and in terms of the production process (Fig. 19).

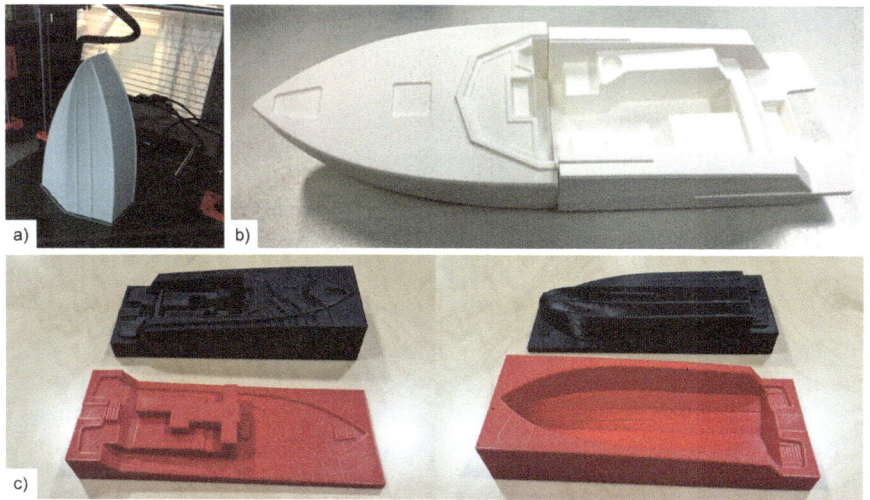

Fig. 17   3D printing by extrusion of fused filament of the designed craft at 1:20 scale (**a** and **b**) and the respective moulds at 1:40 scale (**c**)

Fig. 18 3D printing of the craft's engine using PolyJet technology

Fig. 19 Prototypes produced by additive manufacturing: *Walkaround* version, *Open* version and *Classic* version (from left to right)

The production of prototypes by 3D printing brought competitive advantages, in a faster and less expensive development process than traditional methods (CNC, injection moulding, among others). The prototypes created are faithful to the final craft, as the smallest details are replicated. The use of these technologies made it possible to analyse design errors in more detail, study and optimize the final manufacturing process and quickly collect relevant feedback from the different sectors of the company (design, commercial, and production), which would hardly be achieved only with the visualization of the 3D model on a screen.

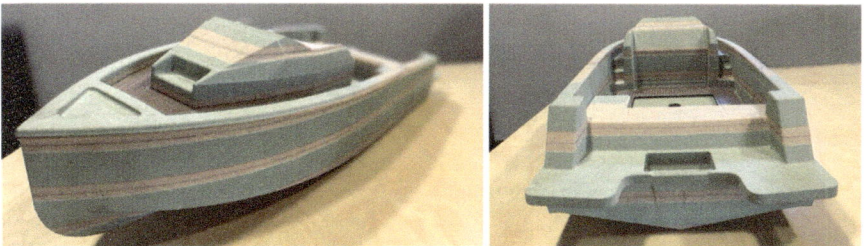

Fig. 20 Unfinished prototype manufactured in MDF by a CNC drilling machine

After the achievement of the final geometry and craft design supported by additive manufacturing, new prototypes at a larger scale (1:10) were manufactured in MDF by a 5 axis CNC drilling machine, for commercial and demonstration purposes in a nautical exhibition fair, thus collecting the reactions of the target audience to the new model (Fig. 20).

6 Production Process

The production process starts with the manufacture of male plugs for the hull and deck, from which female fiberglass moulds reinforced with a steel structure will later be produced. The male plugs will be manufactured in a 5 axis CNC drilling machine, from Optima with a working area of 10 m length, 3.5 m width and 2 m height. These male plugs geometries will be manufactured in polystyrene and polyurethane foam coated with appropriated resin paste from Sika and finally fine machining of the final surfaces to achieve a smooth and accurate surface.

Fiberglass moulds reinforced with a steel structure will be then manufactured from the previous described plugs. These moulds will allow manual fiberglass fabrication as well fiberglass infusion process. The mould is first sprayed with gelcoat, then fiberglass cloth is applied, and then resin is used to saturate or "wet out" the fiberglass. After curing, a hull or a boat part will be obtained (Fig. 21).

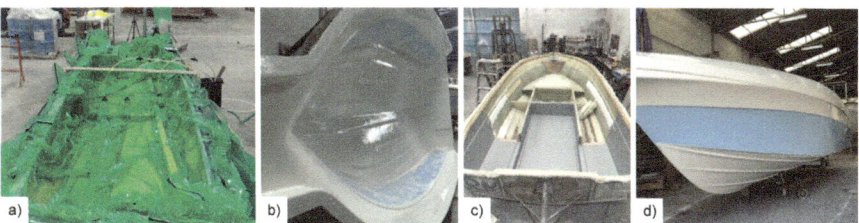

Fig. 21 Craft production process: **a** fiberglass infusion process of a hull; **b** gelcoat sprayed into the mould; **c** fiberglass cloth saturated in resin in the curing process; **d** craft hull after demoulding

7 Conclusions

This paper presents the innovation path for the design and development of a modern recreational craft using advanced technologies of computer-aided design and manufacturing (CAD/CAM), numerical simulation by finite elements (FEM) and additive manufacturing. Design management concepts were analysed from an integrated point of view to assess the internal management and technological capabilities. This analysis allowed to identify the key resources of the company that had to be reinforced as well as the external partnerships needed to accomplish the innovation procedure.

Three-dimensional digital models were created, analysis of static stability of the craft was carried out, FEM simulations were performed to validate the advantages of the two-step hull, and prototypes of different scales were fabricated, allowing the development of a continuous, iterative and flexible creative process, leading to the optimized final solution with the support of the company's internal competences and complementary academic skills.

Through the partnership with University of Porto it was possible to transfer academic/scientific knowledge to the company, leading to a significant technological upgrade. Through this partnership it was possible to shorten the development time of the new model, reduce manufacturing and assembly time, simultaneously designing new production strategies.

This project also made it possible to obtain a clearer and more structured image of the main topics and parameters for design management during the product development process in the context of SMEs, which will allow the development of a future structured methodology, for the implementation of a design management integrated policy in SME's.

Acknowledgements Vitor Carneiro was financed by the Portuguese Foundation for Science and Technology, FCT, (PD/BD/142875/2018) and by the European Social Fund (ESF).

References

Alarcon, J., Lecuona, M., Ormeno, G.: Design management to increase small and medium multi-sector enterprises (SMEs) competitiveness: interdisciplinarity experience with public funding. In: Chova, L.G., Martínez, A.L., Torres, I.C. (eds.) International Conference on Education and New Learning Technologies (Edulearn), pp. 3067–3075. IATED Academy, Barcelona, Spain (2015)

Berends, H., Reymen, I., Stultiëns, R.G.L., Peutz, M.: External designers in product design processes of small manufacturing firms. Des. Stud. **32**, 86–108 (2011). https://doi.org/10.1016/j.destud.2010.06.001

Boats, F.: The History of Recreational Boats. https://www.formulaboats.com/blog/history-of-boating/ (2018). Accessed 10 Oct 2020

Bruce, M., Cooper, R., Vazquez, D.: Effective design management for small businesses. Des. Stud. **20**, 297–315 (1999). https://doi.org/10.1016/S0142-694X(98)00022-2

Carneiro, V., Barata da Rocha, A., Rangel, B., Alves, J.L.: Design management and the SME product development process: a bibliometric analysis and review. She Ji J. Des. Econ. Innov. **7**, 197–222 (2021). https://doi.org/10.1016/j.sheji.2021.03.001

Chiva, R., Alegre, J.: Investment in design and firm performance: the mediating role of design management. J. Prod. Innov. Manag. **26**, 424–440 (2009). https://doi.org/10.1111/j.1540-5885.2009.00669.x

DGRM: Embarcações de Recreio, https://www.dgrm.mm.gov.pt/am-nc-embarcacoes (2018). Accessed 14 Sept 2021

Ghadimi, P., Loni, A., Nowruzi, H., Dashtimanesh, A., Tavakoli, S.: Parametric study of the effects of trim tabs on running trim and resistance of planing hulls. Adv. Shipp. Ocean Eng. **3**, 1–12 (2014)

Hammesfahr, H.: Glass-Cloth or Fabric. US Patent 232,122, 14 Sept 1880

Kobbé, G.: A summary of yachting history. Lotus Mag. **5**, 505–512 (1914)

Marsh, G.: 50 years of reinforced plastic boats. Reinf. Plast. **50**, 16–19 (2006). https://doi.org/10.1016/S0034-3617(06)71125-0

Mitchell, S.: The birth of fiberglass boats. Good Old Boat. **9** (1999)

Oakley, M.: Product design and development in small firms. Des. Stud. **3**, 5–10 (1982). https://doi.org/10.1016/0142-694X(82)90073-4

Oliveira, L., Lopes, A.: Mecânica dos Fluídos. Lidel, Lisboa (2013)

Press Association: Little ships set sail for Dunkirk on 75th anniversary of WWII evacuation. https://www.theguardian.com/world/2015/may/21/dunkirk-operation-dynamo-75th-anniversary-little-ships-second-world-war (2015). Accessed 04 Feb 2022

Rudow, L.: Stepped Hulls vs. Traditional V Bottom: Everything you Need to Know. https://www.boats.com/reviews/stepped-hulls-vs-traditional-v-bottom-everything-you-need-to-know/ (2012). Accessed 22 Jan 2021

Simpson, A.: Understanding the All-Revealing Gz Curves. https://www.sailboat-cruising.com/gz-curves.html (2021). Accessed 14 Nov 2021

Engineering and Manufacturing of Cementitious Mortars with Low Capillary Suction for the Applications in Historical Buildings

Natalia Szemiot⬤, Łukasz Sadowski, and Sławomir Czarnecki

Excessive moisture can cause various types of damage to buildings. The capillary moisture primarily comes from the groundwater. The results obtained during the research allowed to determine the influence of additive (sodium silicate $Na_2O + SiO_2$) in various amounts, as well as the amount of quartz sand in the process of designing a cement masonry mortar, on capillary suction. In the research 3 different types of cement masonry mortar (M5, M10 and M15) and one type of sand (quartz sand) were analysed. First, the capillary suction of the quartz sand was researched. Afterwards, nine cement masonry mortar bars were made. Then these bars were used to research the capillary suction. The results of this test shown that M15 cement masonry mortar with quartz sand and a higher amount of sodium silicate had the most favourable result (37 mm height of capillary suction after 120 h of research). On the other hand, M5 cement masonry mortar without additive had the least favourable result (160 mm height of capillary suction after 24 h of research). The results present that the additive of sodium silicate in the production of cementitious mortar can reduce capillary suction.

N. Szemiot (✉) · Ł. Sadowski · S. Czarnecki
Department of Materials Engineering and Construction Processes, Wroclaw University of Science and Technology, Wrocław, Poland
e-mail: nataliaszemiot@gmail.com

Ł. Sadowski
e-mail: lukasz.sadowski@pwr.edu.pl

S. Czarnecki
e-mail: slawomir.czarnecki@pwr.edu.pl

1 Introduction

Use of buildings and unfavourable heat and moisture conditions lead to the degradation of buildings. The durability of building materials depends on their construction, which, with the participation of moisture, leads to their destruction [1–4]. Capillary suction is a physical phenomenon in which water in a porous material (for example brick) moves upward by capillary forces. Capillary suction of building materials can even cause failure in building [5–8]. With every year, the issue of preserving valuable, historical buildings in good condition (Fig. 1).

Scientists are trying to reduce the capillary suction of cement mortar by using: sodium oleate, powdered silicone, powders, different types of admixtures, the freeze–thaw environment, used of recycled materials [9–13]. However, the solutions presented by scientists are more time-consuming, complicated and expensive than the one proposed in this article. The solution presented in this article is more novel, ecological and optimization of manufacturing processes. Hitherto, there has been no tests on the reduction in the capillary suction rate of cementitious mortars with the use of additive $Na_2O + SiO_2$ (sodium silicate) and quartz sand. Moreover, there has been research on the reduction in the capillary suction rate of cementitious mortars with use (in various proportions) of sodium silicate and basalt sand [14].

The novelty of this research is to design of a cement masonry mortar with low capillary suction with sodium silicate and quartz sand were used. The production of this type of cement mortar is aimed to reduce rate of capillary suction. Furthermore, the aim of the tests was also to show the impact of using sodium silicate in different proportions in the production of cement mortar on capillary suction.

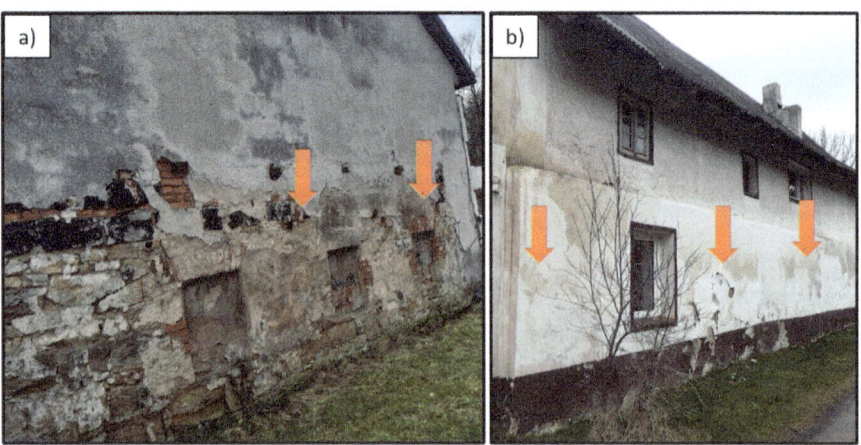

Fig. 1 Examples of moisture and destruction of buildings: **a** destruction of building, **b** moisture line

2   Materials and Methods

This chapter the materials and methods of the research were presented.

2.1   Quartz Sand

In the research, quartz sand with a water absorption $WA_{24}0.6$, a density of $\rho d =$ 2.62 g/cm^3 was used. The grain size of quartz sand ranges from 0.063 to 2.5 mm,

where:

$WA_{24}X$

X   Maximum aggregate water absorption determined after 24 h of being immersed in water [%].

The quartz sand is a natural aggregate of river origin. Figure 2 presents particle size distribution of the quartz sand. The quartz sand mainly consists of SiO_2 (99%) and other (1%).

2.2   Cement

Portland cement is a gray, loose material, produced by grinding cement clinker with gypsum up to 5%. When mixed with water, cement hardens into a stable mass that is not reactive in normal environments. The chemical composition of Portland cement includes many chemical elements, such as: CaO (64%), SiO_2 (19%), SO_3 (6%), Al_2O_3 (6%), FeO (3%), K_2S (1%) and MgO (1%). In the research, Portland Cement CEM I 42.5 R was used. In the Portland Cement, the chloride content (Cl$^-$) is 0.03%, alkali content (Na_2O_{eq}) is 0.7% and sulphate content (SO_3) is 2.8%. Specific surface area of Portland cement is 3655 cm^2/g, specify destiny is 3.16 g/cm^3. The grain size of Portland cement ranges from 0.02 to 0.14 mm.

Fig. 2 Particle size distribution of quartz sand (based on the information provided in [15])

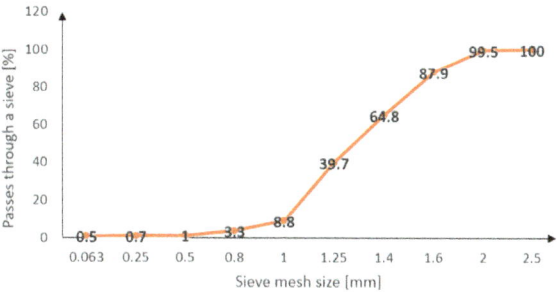

2.3 Sodium Silicate

In the research, sodium silicate was used. Chemical formula of sodium silicate is $Na_2O + SiO_2$. Sodium silicate is a colourless, white or semi-translucent, odourless, viscous liquid. Sodium silicate is made of micelles surrounded by water strongly bound to them, weakly bound water and water not bound to micelles. The density of sodium silicate ranges from 1260 to 1710 kg/m^3. The substance is non-flammable, inorganic and pH of sodium silicate is 11–13 in a temperature of 20 °C. Softening point of sodium silicate is 550–670 °C and pour point is 730–870 °C. Sodium silicate does not decompose at temp. below 1400 °C.

2.4 Research of the Capillary Suction of the Sand

At the beginning, the quartz sand capillary suction test was performed. The test was performed on 3 samples of quartz sand. Each sample of quartz sand was placed in a plexiglass pipe. The dimensions of the tube are shown in the figure below (Fig. 3). The tubes were filled with quartz sand to a height of 1000 mm. Then, the tubes were placed (separately) in a glass vessel filled with water. Then, after 5, 15, 30 min and after 1, 2, 3, 5, 10, 24, 48, 72, 96, 120 h, the height of the capillary suction was measured. Figure 3 presents a research of capillary suction of the quartz sand.

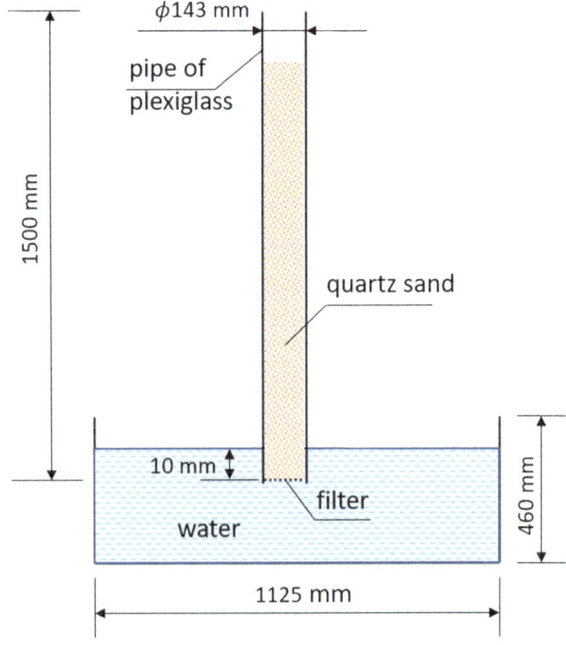

Fig. 3 Test stand for the capillary suction measurements (own elaboration based on [16])

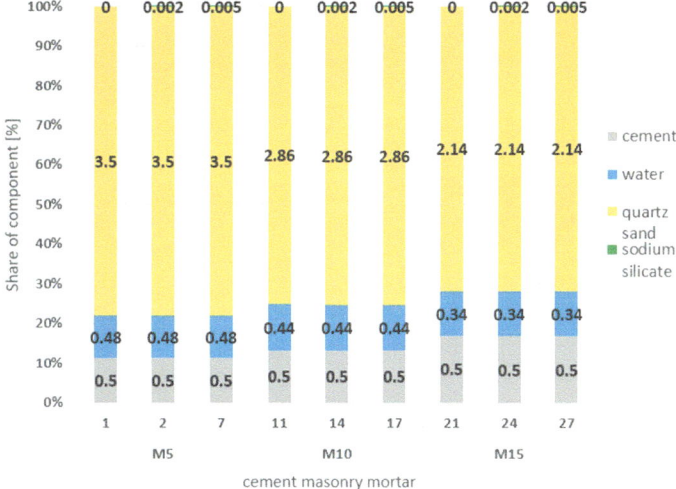

Fig. 4 The proportions of investigated mortars

In the presented research, according to the norm PN-EN 480–1, nine bars made of cement mortar with the following dimensions were prepared: 40 × 40 × 160 [mm]. Figure 4 shows the amounts of the components of cementitious mortars were used. A capillary suction test was performed after 28 days. Then, after 5, 15, 30 min and after 1, 2, 3, 5, 10, 24, 48, 72, 96, 120 h, the height of the capillary suction was measured. The amounts of the components of the cementitious mortar are shown in kilograms.

2.5 *Research of the Capillary Suction of the Cementitious Mortar*

In the research, the cement mortar bars with dimensions 40 × 40 × 160 [mm] were transferred to a glass vessel and embedded in water to the height of 10 mm. After 5, 15, 30 min and after 1, 2, 3, 5, 10, 24, 48, 72, 96, 120 h, the indexes of the capillary suction were measured.

Figure 5a presents the stand for the capillary suction measurements of cement masonry mortar bar. Figure 5b presents M5 and M15 cementitious mortar bars.

3 Results

This chapter the results of the research were analysed.

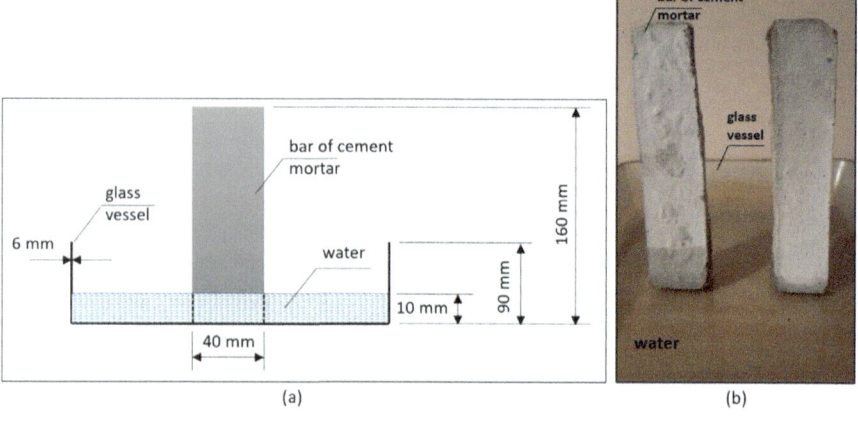

Fig. 5 The stand for the capillary suction measurements of cement masonry mortar bar: **a** scheme, **b** M5 and M15 cement masonry mortar bars embedded in water

3.1   The Capillary Suction of the Quartz Sand

Three samples to test the capillary suction of quartz sand were used. Each of the three samples 13 times was measured, then the mean was calculated. All results are presented in Table 1. The results of capillary suction of quartz sand samples were similar to each other.

Table 1   The capillary suction test of the quartz sand

Quartz				
	h (mm)			
Time	Sample 1	Sample 2	Sample 3	Average
5 min	49	47	49	48.33
15 min	104	104	106	104.67
30 min	151	150	152	151.00
1 h	172	171	174	172.33
2 h	209	205	209	207.67
3 h	220	219	222	220.33
5 h	235	236	238	236.33
10 h	261	262	263	262.00
24 h	283	281	283	282.33
48 h	291	296	298	295.00
72 h	298	299	301	299.33
96 h	300	301	305	302.00
120 h	307	305	308	306.67

Fig. 6 Line chart of the capillary suction of quartz sand and basalt sand—comparison

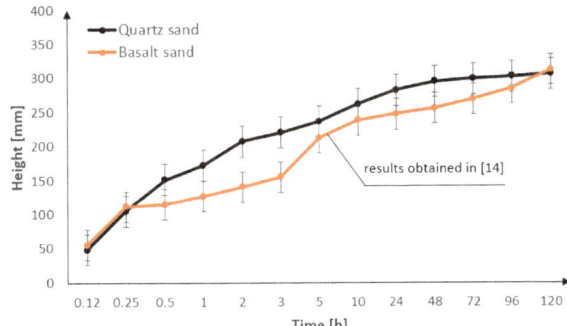

Figure 6 presents the line chart of the capillary suction of quartz and basalt sand. In the first 15 min of the test, the values of the capillary suction index of quartz and basalt sand was similar. From 15 min to about 100 h of the test, the capillary suction of quartz sand was on average 17% higher than the capillary suction of basalt sand. Finally, in the last hour of the test, the quartz sand had a lower capillary suction index, compared to the capillary suction of the basalt sand. The quartz sand was 2% more resistant to capillary suction than the basalt sand.

3.2   The Capillary Suction of the Cementitious Mortars

Figure 7 presents the capillary suction of the bar of M5 cement masonry mortar. Two bars of M5 cement mortar were tested (the composition of the cement bars is shown in Fig. 4). In this research, each bar was measured 13 times. Figure 7a shows a comparison of the capillary suction of two M5 cement mortar bars with a lower amount of sodium silicate: quartz sand was added to the first cement mortar, and basalt sand was added to the second. The bar of M5 cementitious mortar with quartz sand had more favourable result, in compared to the cement mortar bar with basalt sand. Figure 7b shows a comparison of the capillary suction of two M5 cement mortar bars with a higher amount of sodium silicate: quartz sand was added to the first cement mortar, and basalt sand was added to the second. The bar of M5 cement mortar with quartz sand had more favourable result, in compared to the cement mortar bar with basalt sand. However, the M5 cement mortar with a greater proportion of additive had more favourable result than the M5 cement mortar with a lower proportion of sodium silicate, regardless of the type of sand used.

Figure 8 presents the capillary suction of the bar of M10 cement masonry mortar. Two bars of M10 cement mortar were tested (the composition of the cement bars is shown in Fig. 4). In the research, each bar was measured 13 times. Figure 8a shows a comparison of the capillary suction of two M10 cement mortar bars with a lower amount of sodium silicate: quartz sand was added to the first cement mortar, and basalt sand was added to the second. The bar of M10 cement mortar with quartz sand

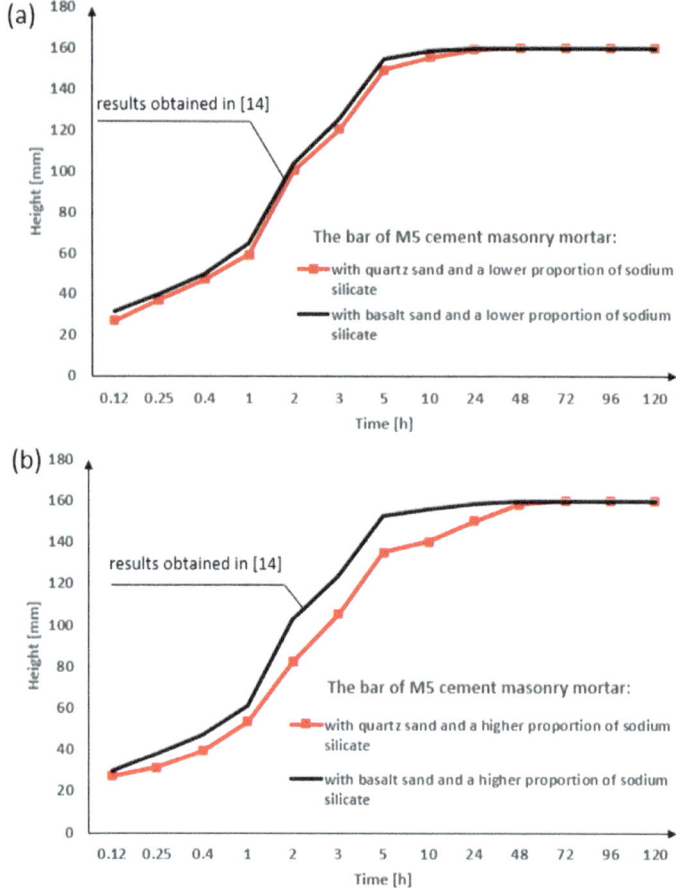

Fig. 7 Line chart of capillary suction of M5 cementitious mortar bars with quartz sand and basalt sand: **a** with a lower proportion of additive, **b** with a higher proportion of additive

had more favourable result, in compared to the cement mortar bar with basalt sand. Figure 8b shows a comparison of the capillary suction of two M10 cement mortar bars with a higher amount of sodium silicate: quartz sand was added to the first cement mortar, and basalt sand was added to the second. The bar of M10 cement mortar with quartz sand had more favourable result, in compared to the cement mortar bar with basalt sand. However, the M10 cement mortar with a greater proportion of additive had more favourable result than the M10 cement mortar with a lower proportion of sodium silicate, regardless of the type of sand used. On the other hand, the bar of M5 cement mortars had a higher capillary suction than the M10 cement mortar bars. For the production of the M5 cementitious mortar, more sand was used than in the M10 cementitious mortar. In the last hour of the test, the capillary suction of the M10 cement mortar with quartz sand and a lower proportion of sodium silicate is about

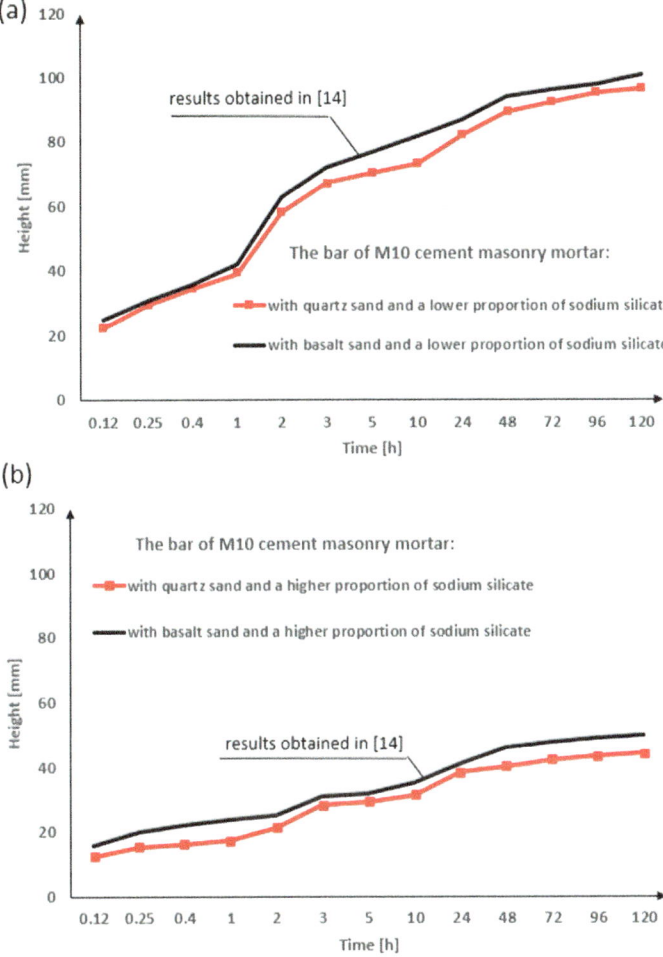

Fig. 8 Line chart of capillary suction of M10 cementitious mortar bars with quartz sand and basalt sand: **a** with a lower proportion of additive, **b** with a higher proportion of additive

54% higher than the capillary suction of the M10 cement mortar with quartz sand and a higher proportion of sodium silicate. However, in the last hour of the test, the capillary suction of M10 cement mortar with basalt sand and a higher proportion of sodium silicate is about 12% higher than the capillary suction of the M10 cement mortar with quartz sand and a higher proportion of sodium silicate.

Figure 9 presents the capillary suction of the bar of M15 cement masonry mortar. Two bars of M15 cement mortar were tested (the composition of the cement bars is shown in Fig. 4). In the research, each bar was measured 13 times. Figure 9a shows a comparison of the capillary suction of two M15 cement mortar bars with a lower amount of sodium silicate: quartz sand was added to the first cement mortar, and basalt sand was added to the second. The bar of M15 cement mortar with quartz sand had more favourable result, in compared to the cement mortar bar with basalt sand. Figure 9b shows a comparison of the capillary suction of two bars of M15 cementitious mortar with a higher amount of sodium silicate: quartz sand was added to the first cement mortar, and basalt sand was added to the second. The bar of M15 cement mortar with quartz sand had more favourable result, in compared to the cement mortar bar with basalt sand. However, the M15 cementitious mortar with a greater proportion of additive had more favourable result than the M15 cementitious mortar with a lower proportion of additive, regardless of the type of sand used. On the other hand, the bar of M15 cementitious mortar with quartz sand and a greater proportion of additive had more favourable result than the other bars. Tests of the capillary suction presented that the M15 cement mortar, in comparison to the M5 and M10 cement mortars, had the most favourable result. The reason for these results was the amount of the components that were used in the cementitious mortar when compared to the other cementitious mortars. The smallest amount of quartz sand was used in the production of M15 cement mortar. In the last hour of the test, the capillary suction of the M15 cement mortar with quartz sand and a lower proportion of sodium silicate is about 33% higher than the capillary suction of the M15 cement mortar with quartz sand and a higher proportion of sodium silicate. However, in the last hour of the test, the capillary suction of M15 cement mortar with basalt sand and a higher proportion of sodium silicate is about 5% higher than the capillary suction of the M15 cement mortar with quartz sand and a higher proportion of sodium silicate.

Figure 10a shows the capillary suction of quartz and basalt sand and the capillary suction of the bar of M5 cement mortar with the quartz sand without sodium silicate and the M5 cement mortar bar with the basalt sand without sodium silicate. Figure 10b shows the capillary suction of quartz and basalt sand and the capillary suction of the bar of M10 cementitious mortar without sodium silicate with the quartz sand and the M10 cementitious mortar bar without sodium silicate with the basalt sand. Figure 10c presents the capillary suction of quartz and basalt sand, as well as the results of the capillary suction of the bar of M15 cementitious mortar without sodium silicate with the quartz sand and the M15 cement mortar bar without sodium silicate with the basalt sand. The quartz and basalt sand had the least favourable result, and their values definitely differ from the cementitious mortar bars. Moreover, line charts show that the higher the type of cementitious mortar (the less sand used), the lower the capillary suction. From 1 to 10 h of test, the capillary suction of the M5 cement mortar grows faster compared to the M10 and M15 mortars. The capillary suction of M5 cement mortar is about 40% higher than the capillary suction of M15 cement mortar.

4   Conclusions

In this research, the main aim is to find a way to produce a cement masonry mortar with low capillary suction. For this purpose, the article presents the impact of the

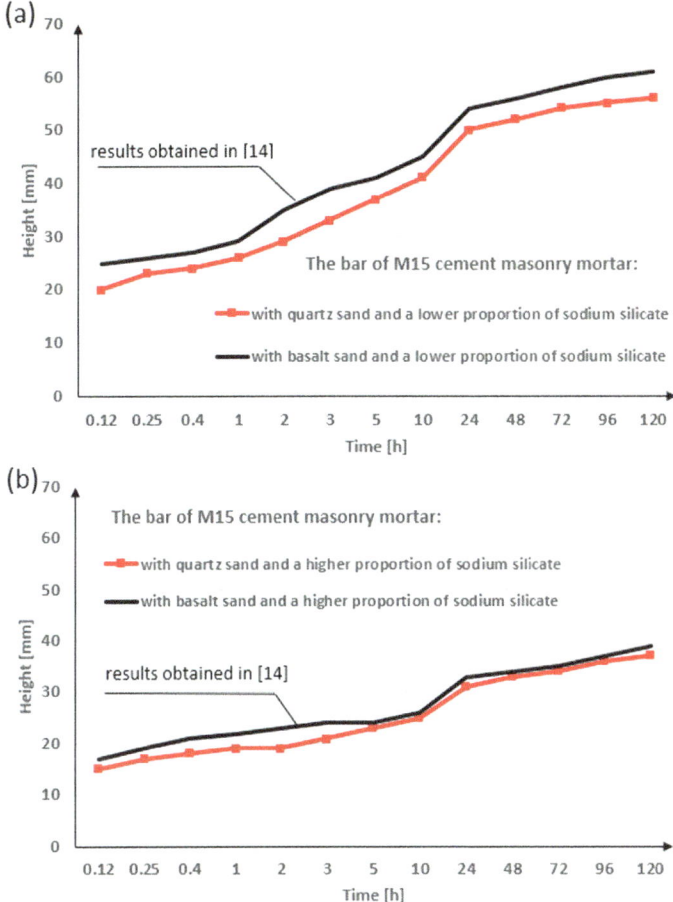

Fig. 9 Line chart of capillary suction of M15 cementitious mortar bars with quartz sand and basalt sand: **a** with a lower proportion of additive, **b** with a higher proportion of additive

sodium silicate (in various proportions) additive on the capillary suction index of the cement mortar. In this research, M5, M10 and M15 cementitious mortars and quartz sand were used. It was observed that sodium silicate has positive influence on a capillary suction index of cement masonry mortar. The researched cement mortars differed from each other in the amount of sand and water used for the production. The amount of cement used was the same in each type of cement mortar.

Furthermore, the following observations can be stated:

- The M15 cement masonry mortar with the quartz sand and more amount of additive (sodium silicate) had the lowest index of capillary suction. The addition of sodium silicate to the cement mortar reduces the capillary suction of the mortar. The amount of sand added to the cement masonry mortar is also important. For

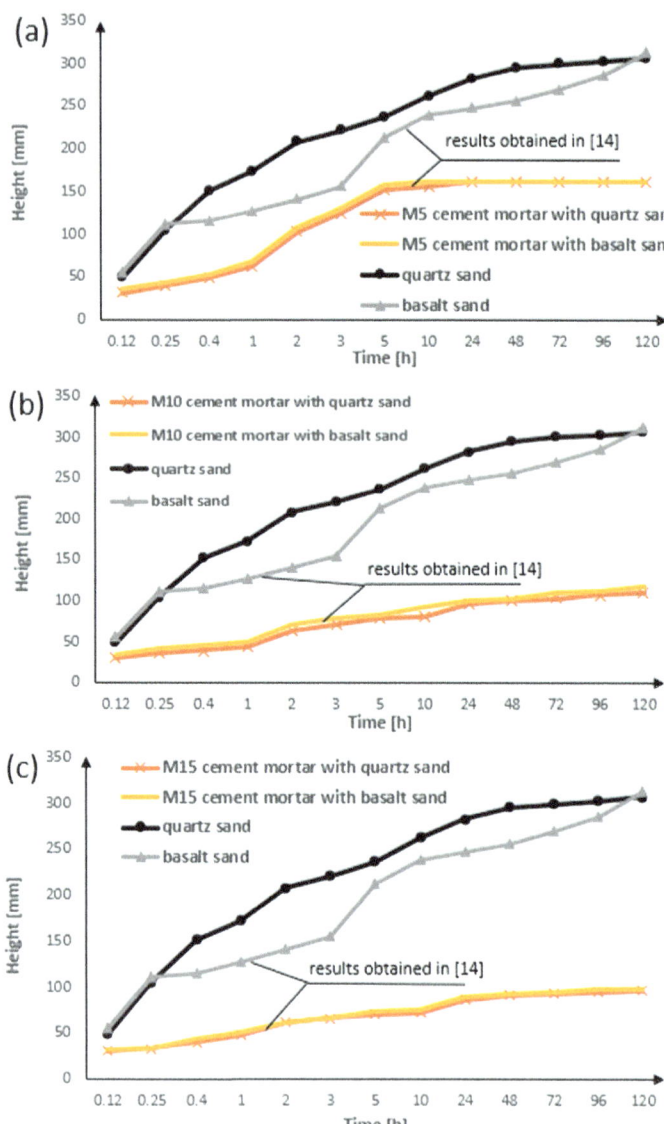

Fig. 10 Comparison of the capillary suction of the quartz sand and basalt sand with: **a** M5 cement mortar bar without sodium silicate, **b** M10 cement mortars bar without sodium silicate, **c** M15 cement mortar bars without sodium silicate

the production of the M15 cementitious mortar, in comparison to the M5 and M10 cement mortars, the smallest amount of quartz sand was used.

- The M5 cement masonry mortar, in comparison to the M10 and M15 cement masonry mortars, had the highest index of capillary suction. For the production of the M5 cementitious mortar, in comparison to the M10 and M15 cement mortars, the highest amount of quartz sand was used, which resulted in a higher capillary suction index.
- The addition of sodium silicate to cement masonry mortar causes a reduction rate of capillary suction of cement mortar.

It should be stated that sodium silicate has great potential to be used as additive to cement masonry mortar. The perspective for the further research is the use of sodium silicate as an additive to cement masonry mortar and the use of other types of sand (e.g. granite sand).

References

1. Hoła, B., Hoła, A.: The latest scientific problems related to the implementation and diagnostics of construction objects. Appl. Sci. **11**(13), 6184 (2021)
2. Öchsner, A., Murch, G., Shokuhfar, A., Delgado, J.: Salt degradation in stone of old buildings. Defect Diffus. Forum **2**, 337–342 (2013)
3. Kłosowski, G., Hoła, A., Rymarczyk, T., Skowron, Ł., Wołowiec, T., Kowalski, M.: The concept of using LSTM to detect moisture in brick walls by means of electrical impedance tomography. Energies **14**(22), 11 (2021)
4. Hoła, A., Sadowski, Ł.: Verification of a nondestructive method for assessing the humidity of saline brick walls in historical buildings. Appl. Sci. **10**(19) (2020)
5. Hall, C., Yau, R.: Water movement in porous building materials—IX. The water absorption and sorptivity of concretes. Build. Environ. **22**, 77–82 (1987)
6. Raimondo, M., Dondi, M., Gardini, D., Guarini, G., Mazzanti, F.: Predicting the initial rate of water absorption in clay bricks. Constr. Build. Mater. **23**(7), 2623–2630 (2009)
7. Wang, Y., Wang, W., Wang, D., Liu, Y., Liu, J.: Study on the influence of sample size and test conditions on the capillary water absorption of porous building materials. J. Build. Eng. **43** (2021)
8. Hall, C.: Water movement in porous building materials—IV. The initial surface absorption and the sorptivity. Build. Environ. **16**, 201–207 (1981)
9. Lagazzo, A., Vicini, S., Cattaneo, C., Botter, R.: Effect of fatty acid soap on microstructure of lime-cement mortar **116**, 384–390 (2016)
10. Gao, Q., Ma, Z., Jianzhuang, X., Li, F.: Effects of imposed damage on the capillary water absorption of recycled aggregate concrete. Adv. Mater. Sci. Eng. (2018)
11. Deboucha, W., Leklou, N., Khelidj, A., Oudjit, M.: Natural pozzolana addition effect on compressive strength and capillary water absorption of Mortar. Energy Procedia **139**, 689–695 (2017)
12. Oltulu, M., Sahin, R.: Effect of nano-SiO_2, nano-Al_2O_3 and nano-Fe_2O_3 powders on compressive strengths and capillary water absorption of cement mortar containing fly ash: a comparative study. Energy Build. **58**, 292–301 (2013)
13. Guardia, C., Schicchi, D., Caggiano, A., Barluenga, G., Koenders, E.: On the capillary water absorption of cement-lime mortars containing phase change materials: experiments and simulations. Build. Simul. **13**, 19–31 (2020)

14. Szemiot, N., Sadowski, Ł: The design of cement mortar with low capillary suction: understanding the effect of fine aggregate and sodium silicate. Materials **15**, 1517 (2022)
15. Szemiot, N.: Wpływ domieszki szkła wodnego na podciąganie kapilarne zapraw cementowych. Materiały Budowlane **11**, 39–40 (2021)
16. Wysocka, M., Tymosiak, D., Szypcio, Z.: Prędkość wznoszenia kapilarnego w gruntach niespoistych. Budownictwo i Inżynieria Środowiska **4**, 167–172 (2013)

Printed by Printforce, the Netherlands